Supercritical Fluid Methods and Protocols

METHODS IN BIOTECHNOLOGY™

John M. Walker, *SERIES EDITOR*

13. **Supercritical Fluid Methods and Protocols,** edited by *John R. Williams and Anthony A. Clifford, 2000*
12. **Environmental Monitoring of Bacteria,** edited by *Clive Edwards, 1999*
11. **Aqueous Two-Phase Systems,** edited by *Rajni Hatti-Kaul, 1999*
10. **Carbohydrate Biotechnology Protocols,** edited by *Christopher Bucke, 1999*
9. **Downstream Processing Methods,** edited by *Mohamed A. Desai, 2000*
8. **Animal Cell Biotechnology,** edited by *Nigel Jenkins, 1999*
7. **Affinity Biosensors: *Techniques and Protocols,*** edited by *Kim R. Rogers and Ashok Mulchandani, 1998*
6. **Enzyme and Microbial Biosensors: *Techniques and Protocols,*** edited by *Ashok Mulchandani and Kim R. Rogers, 1998*
5. **Biopesticides: Use and Delivery,** edited by *Franklin R. Hall and Julius J. Menn, 1998*
4. **Natural Products Isolation,** edited by *Richard J. P. Cannell, 1998*
3. **Recombinant Proteins from Plants: *Production and Isolation of Clinically Useful Compounds,*** edited by *Charles Cunningham and Andrew J. R. Porter, 1998*
2. **Bioremediation Protocols ,** edited by *David Sheehan, 1997*
1. **Immobilization of Enzymes and Cells,** edited by *Gordon F. Bickerstaff, 1997*

METHODS IN BIOTECHNOLOGY™

Supercritical Fluid Methods and Protocols

Edited by

John R. Williams

College of Science, Sultan Qaboos University, Sultanate of Oman

and

Anthony A. Clifford

School of Chemistry, Leeds, UK

Humana Press Totowa, New Jersey

Softcover reprint of the hardcover 1st edition 2000
999 Riverview Drive, Suite 208
Totowa, New Jersey 07512

This publication is printed on acid-free paper. ∞
ANSI Z39.48-1984 (American Standards Institute) Permanence of Paper for Printed Library Materials.

Cover design by Patricia F. Cleary.

For additional copies, pricing for bulk purchases, and/or information about other Humana titles, contact Humana at the above address or at any of the following numbers: Tel: 973-256-1699; Fax: 973-256-8341; E-mail: humana@humanapr.com, or visit our Website at www.humanapress.com

10 9 8 7 6 5 4 3 2 1

Library of Congress Cataloging in Publication Data

Supercritical fluid methods and protocols / edited by John R. Williams and Anthony A. Clifford.
p. cm. -- (Methods in biotechnology ; 13)
Includes bibliographical references and index.
ISBN 978-1-61737-079-3 e-ISBN 978-1-59259-030-8
1. Supercritical fluid extraction--Laboratory manuals. 2. Supercritical fluid chromatography--Laboratory manuals. 3. Biomolecules--Separation--Laboratory manuals. I. Williams, John R., 1967- II. Clifford, Tony. III. Series.

QP519.9.S85 S87 2001
572'.36'028--dc21

00-024567

Preface

Over the last 15 years, there has been renewed interest in supercritical fluids owing to their unique properties and relatively low environmental impact. Greatest attention has been given to the extraction and separation of organic compounds. Supercritical fluids have also been successfully used for particle production, as reaction media, and for the destruction of toxic waste. Supercritical carbon dioxide has been the most widely used supercritical fluid, mainly because it is cheap, relatively nontoxic, and has convenient critical values. Supercritical fluids have also been used on analytical and preparative scales for many biological and other applications.

Many papers have been published on the use of supercritical fluids. However, few have acted as a detailed instruction manual for those wanting to use the techniques for the first time. We anticipate that this *Methods in Biotechnology* volume, *Supercritical Fluid Methods and Protocols* will satisfy the need for such a book.

Every chapter has been written by experienced workers and should, if closely followed, enable workers with some or no previous experience of supercritical fluids to conduct experiments successfully at the first attempt. The Introduction to each chapter gives the reader all the necessary background information. The Materials and Methods sections describe, in detail, the apparatus and steps needed to complete the protocol quickly, with a minimum of fuss. The Notes section, an acclaimed feature of the *Methods in Biotechnology* series, gives additional information not normally seen in published papers that enable the procedures to be conducted easily. Some of the chapters describe how the procedures can be modified for application to new situations. The first chapter is not a detailed procedure, but a theoretical, general introduction to the area of supercritical fluids intended to instruct novices in this branch of technology.

It is envisaged that *Supercritical Fluid Methods and Protocols* will be useful to both student and experienced research workers in biology and related areas. Our hope is that the experience gained when using these techniques will give these workers the confidence to explore new applications for supercritical fluids.

One can envisage a time in the future when the use of sub- and supercritical carbon dioxide and water becomes very important in laboratory work, with organic solvent use considerably reduced.

Finally, we would like to thank Professor John Walker for allowing us to edit this volume and for his cooperation during the compiling of this book. We would also like to acknowledge Professor E. D. Morgan of Keele University, UK for passing this opportunity on to us. We thank Thomas Lanigan and his colleagues at Humana for their help in seeing our book through press.

John R. Williams
Anthony A. Clifford

Contents

Contributors

PAOLO ALESSI • *Dipartimento di Ingegneria Chimica, dell'Ambiente e delle Materie Prime, University of Trieste, Trieste, Italy*

KEIICHI AOE • *Analytical Research Laboratory, Tanabe Seiyaku Co., Ltd., Osaka, Japan*

VIDDESH R. BARI • *Department of Chemistry, Ismail Yusuf College of Arts, Commerce and Science, Mumbai, India*

SUSANA BARREIROS • *Instituto de Tecnologia Química e Biológica, Universidade Nova de Lisboa, Oeiras, Portugal*

JOSÉ L. BERNAL • *Department of Analytical Chemistry, Faculty of Sciences, University of Valladolid, Valladolid, Spain*

ALBERTO BERTUCCO • *Istituto di Impianti Chimici, University of Padova, Padova, Italy*

CHRISTINA BORCH-JENSEN • *Department of Chemical Engineering, Technical University of Denmark, Lyngby, Denmark*

DOUGLAS BRAY • *Department of Biological Sciences, University of Lethbridge, Lethbridge, Canada*

FABIO CARLI • *Vectorpharma SPA, Trieste, Italy*

ANTHONY A. CLIFFORD • *School of Chemistry, University of Leeds, Leeds, UK*

PHILIPPE CHRISTEN • *Laboratoire de Chimie Analytique Pharmaceutique, Université de Genève, Genève, Switzerland*

MICHAEL D. COLE • *Forensic Science Unit, University of Strathclyde, Glasgow, UK*

MARIA L. COLOMBO • *Institute of Pharmacological Science, University of Milan, Milan, Italy*

TERESA CORRÊA DE SAMPAIO • *Instituto de Tecnologia Química e Biológica, Universidade Nova de Lisboa, Oeiras, Portugal*

ANGELO CORTESI • *Dipartimento di Ingegneria Chimica, dell'Ambiente e delle Materie Prime, University of Trieste, Trieste, Italy*

JOHN R. DEAN • *School of Applied and Molecular Sciences, University of Northumbria, Newcastle upon Tyne, UK*

MARÍA J. DEL NOZAL • *Department of Analytical Chemistry, Faculty of Sciences, University of Valladolid, Valladolid, Spain*

Urmila J. Dhorda • *Department of Chemistry, Ismail Yusuf College of Arts, Commerce and Science, Mumbai, India*
Walter Fiddler • *Agricultural Research Service, Eastern Regional Research Center, US Department of Agriculture, Wyndmoor, PA*
Brenda S. Fuerst • *Food Safety Inspection Service, Midwestern Laboratory, US Department of Agriculture, St. Louis, MO*
Mário S. Galhiane • *Institute of Chemistry of São Carlos, University of São Paulo, São Carlos, Brazil*
Thomas Gamse • *Institut für Thermische Verfahrenstechnik und Umwelttechnik, Technische Universität Graz,Graz, Austria*
Michael O. Gaylor • *Department of Environmental Sciences, Virginia Institute of Marine Sciences, College of William and Mary, Gloucester Point, VA*
Ricky L. Green • *Express Separations Limited, Leeds, UK*
Werner Haerdi • *Laboratoire de Chimie Analytique Pharmaceutique, Université de Genève, Pavillon des Isotopes, Genève, Switzerland*
Robert C. Hale • *Department of Environmental Sciences, Virginia Institute of Marine Sciences, College of William and Mary, Gloucester Point, VA*
Kari Hartonen • *Laboratory of Analytical Chemistry, Department of Chemistry, University of Helsinki, Helsinki, Finland*
Carolyn Henry • *Midwestern Laboratory, Food Safety Inspection Service, US Department of Agriculture, St. Louis, MO*
Rainer Huopalahti • *Department of Biochemistry and Food Chemistry, University of Turku, Turku, Finland*
Brooks M. Hybertson • *Webb-Waring Institute for Cancer, Aging and Antioxidant Research, University of Colorado Health Sciences Center, Denver, CO*
Eila P. Järvenpää • *Department of Biochemistry and Food Chemistry, University of Turku, Turku, Finland*
Juan J. Jiménez • *Department of Analytical Chemistry, Faculty of Sciences, University of Valladolid, Valladolid, Spain*
Ireneo Kikic • *Dipartimento di Ingegneria Chimica, dell'Ambiente e delle Materie Prime, University of Trieste, Trieste, Italy*
Pascal Kintz • *Institut de Médecine Légale, Cedex, France*
Marcel Kohler • *Laboratoire de Chimie Analytique Pharmaceutique, Université de Genève, Pavillon des Isotopes, Genève, Switzerland*
Fernando M. Lanças • *Institute of Chemistry of São Carlos, University of São Paulo, São Carlos, Brazil*
Keith L. Levien • *Department of Chemical Engineering, Oregon State University, Corvallis, OR*

ALAN R. LIGHTFIELD • *Eastern Regional Research Center, Agricultural Research Service, US Department of Agriculture, Wyndmoor, PA*

BEN LIU • *Department of Pharmacy, Hubei College of Traditional Chinese Medicine, People's Republic of China*

EDWIN LUDKIN • *School of Applied and Molecular Sciences, University of Northumbria, Ellison Building, Newcastle upon Tyne, UK*

MAGNUS MAGNUSSEN • *Food and Environmental Institute, Thorshavn, Faroe Islands*

KARIN E. MARKIDES • *Department of Analytical Chemistry, Uppsala University, Uppsala, Sweden*

ROLF MARR • *Institut für Thermische Verfahrenstechnik und Umwelttechnik, Technische Universität Graz, Graz, Austria*

MARÍA T. MARTÍN • *Department of Analytical Chemistry, Faculty of Sciences, University of Valladolid, Valladolid, Spain*

ROBERT J. MAXWELL • *Eastern Regional Research Center, Agricultural Research Service, US Department of Agriculture, Wyndmoor, PA*

HARALD MICHOR • *Institut für Thermische Verfahrenstechnik und Umwelttechnik, Technische Universität Graz, Graz, Austria*

BRIAN MINTY • *School of Applied Sciences, University of Glamorgan, Glamorgan, UK*

JØRGEN MOLLERUP • *Department of Chemical Engineering, Technical University of Denmark, Lyngby, Denmark*

FUJIO MORISHITA • *Department of Material Chemistry, Graduate School of Engineering, Kyoto University, Kyoto, Japan*

JEFFREY J. MORRELL • *Department of Forest Products, Oregon State University, Corvallis, OR*

ANDREA MOSSA • *Institute of Pharmacological Science, University of Milan, Milan, Italy*

NORIYUKI NISHIMURA • *Analytical Research Laboratory, Tanabe Seiyaku Co., Ltd., Osaka, Japan*

HIDETAKA NORITOMI • *Department of Applied Chemistry, Graduate School of Engineering, Tokyo Metropolitan University, Tokyo, Japan*

PAOLO PALLADO • *Exenia Group srl., Albignasego, Italy*

JOHN W. PENSABENE • *Eastern Regional Research Center, Agricultural Research Service, US Department of Agriculture, Wyndmoor, PA*

OWEN W. PARKS • *Eastern Regional Research Center, Agricultural Research Service, US Department of Agriculture, Wyndmoor, PA*

EDWARD D. RAMSEY • *School of Applied Sciences, University of Glamorgan, Glamorgan, UK*

ANTHONY T. REES • *Nycomed Amersham, Cardiff Laboratories, Cardiff, UK*

MARJA-LIISA RIEKKOLA • *Laboratory of Analytical Chemistry, Department of Chemistry, University of Helsinki, Helsinki, Finland*
SANDRA R. RISSATO • *Institute of Chemistry of São Carlos, University of São Paulo, São Carlos, Brazil*
TADASHI SATO • *Analytical Research Laboratory, Tanabe Seiyaku Co., Ltd., Osaka, Japan*
HELMUT SCHWAB • *Institut für Biotechnologie, Technische Univerität Graz, Graz, Austria*
FRANCISCO J. SEÑORÁNS • *Ciencia y Tecnologia de Alimentos, Facultad de Ciencias, Universidad Autónoma de Madrid, Madrid, Spain*
MARIA L. SERNA • *Department of Analytical Chemistry, Faculty of Sciences, University of Valladolid, Valladolid, Spain*
ROXANNE J. SHADWELL • *Eastern Regional Research Center, Agricultural Research Service, US Department of Agriculture, Wyndmoor, PA*
CHRISTIAN STAUB • *Institut de Médecine Légale, Genève, Switzerland*
G. KIM STEARMAN • *Center for the Management, Utilization and Protection of Water Resources, Tennessee Technological University, Cookeville, TN*
M. SUNDARESAN • *Department of Chemistry, C.B. Patel Research Centre for Chemistry and Biological Sciences, Mumbai, India*
LAURA TORIBIO • *Department of Analytical Chemistry, Faculty of Sciences, University of Valladolid, Valladolid, Spain*
JEAN-LUC VEUTHEY • *Laboratoire de Chimie Analytique Pharmaceutique, Université de Genève, Pavillon des Isotopes, Genève, Switzerland*
JOHN R. WILLIAMS • *Department of Chemistry, College of Science, Sultan Qaboos University, Sultanate of Oman*
KOJI YAKU • *Analytical Research Laboratory, Tanabe Seiyaku Co., Ltd., Osaka, Japan*

1

Introduction to Supercritical Fluids and Their Applications

Anthony A. Clifford and John R. Williams

1. Pure Substances as Supercritical Fluids

Cagniard de la Tour showed in 1822 that there is a *critical temperature* above which a single substance can only exist as a fluid and not as either a liquid or gas. He heated substances, present as both liquid and vapor, in a sealed cannon, which he rocked back and forth and discovered that, at a certain temperature, the splashing ceased. Later, he constructed a glass apparatus in which the phenomenon could be more directly observed. These phenomena can be put into context by reference to **Fig. 1**, which is a phase diagram of a single substance. The diagram is schematic, the pressure axis is nonlinear, and the solid phase at high temperatures occurs at very high pressures. Further solid phases and also liquid crystal phases can also occur on a phase diagram. The areas where the substance exists as a single solid, liquid, or gas phase are labeled, as is the triple point where the three phases coexist. The curves represent coexistence between two of the phases. If we move upward along the gas–liquid coexistence curve, which is a plot of vapor pressure vs temperature, both temperature and pressure increase. The liquid becomes less dense because of thermal expansion, and the gas becomes more dense as the pressure rises. At the critical point, the densities of the two phases become identical, the distinction between the gas and the liquid disappears, and the curve comes to an end at the *critical point*. The substance is now described as a fluid. The critical point has pressure and temperature co-ordinates on the phase diagram, which are referred to as the critical temperature, T_c, and the critical pressure, p_c, and which have particular values for particular substances, as shown by example in **Table 1** *(1)*.

In recent years, fluids have been exploited above their critical temperatures and pressures, and the term *supercritical fluids* has been used to describe these

From: *Methods in Biotechnology, Vol. 13: Supercritical Fluid Methods and Protocols*
Edited by: J. R. Williams and A. A. Clifford © Humana Press Inc., Totowa, NJ

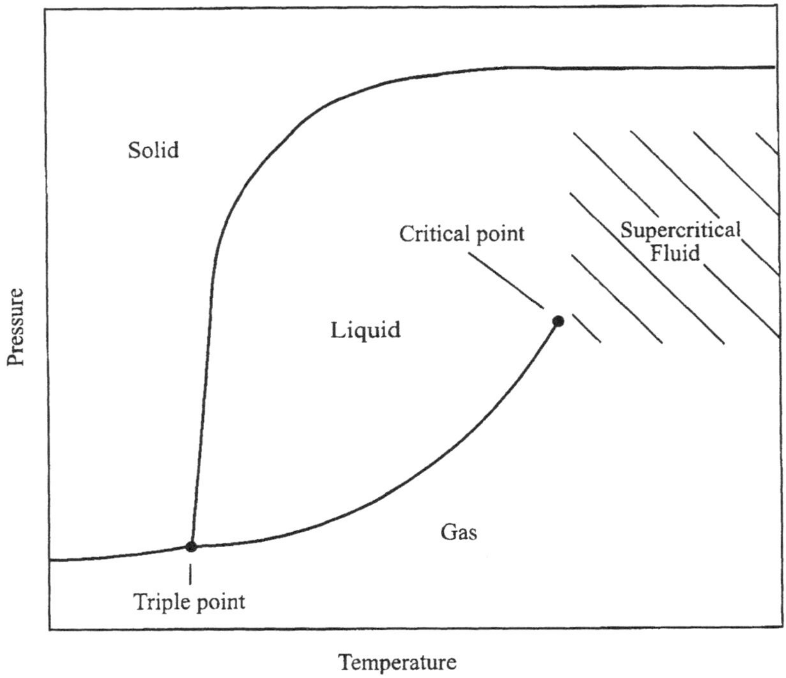

Fig. 1. The phase diagram of a single substance.

media. The greatest advantages of supercritical fluids are realized when they are used not too far above (say within 100 K of) their critical temperatures. Nitrogen gas in a cylinder is a fluid, but is not usually considered as a supercritical fluid, but more often described by an older term as a permanent gas. The region for supercritical fluids is the hatched area in **Fig. 1**. It has been shown to include a region a little below the critical pressure, as processes at these conditions are sometimes included in discussions as "supercritical." Lower pressures are important in practice also because these conditions are relevant to separation stages in supercritical processes. There are no phase boundaries below and to the left of the supercritical region in **Fig. 1**, and behavior does not change dramatically on moving out of the hatched area in these directions. The liquid region to the left of the supercritical region has many of the characteristics of supercritical fluids and is exploited in a similar way. For this reason some people prefer the term *near-critical fluids* and the adjective *subcritical* is also used. The term supercritical fluid has, however, gained currency; is convenient and not problematic provided the definition is not too rigid. Supercritical fluids exhibit important characteristics, such as compressibility, homogeneity, and a continuous change from gaslike to liquidlike prop-

Table 1
Substances Useful as Supercritical Fluids

Substance	Critical Temperature, T_c (K)	Critical Pressure, p_c (bar)
Carbon dioxide	304	74
Water	647	221
Ethane	305	49
Ethene	282	50
Propane	370	43
Xenon	290	58
Ammonia	406	114
Nitrous oxide	310	72
Fluoroform	299	49

Parameters from Reid et al. *(1)*.

erties. These properties are characteristic of conditions inside the hatched area in **Fig. 1** and, to different degrees, in the area around it.

Table 1 shows the critical parameters of some of the important compounds useful as supercritical fluids. One compound, carbon dioxide, has so far been the most widely used because of its convenient critical temperature, cheapness, chemical stability, nonflammability, stability in radioactive applications, and nontoxicity. Large amounts of carbon dioxide released accidentally could constitute a working hazard, given its tendency to blanket the ground, but hazard detectors are available. It is an environmentally friendly substitute for organic solvents. The carbon dioxide is obtained in large quantities as a by-product of fermentation, combustion, and ammonia synthesis and would be released into the atmosphere sooner rather than later, if it were not used as a supercritical fluid. Its polar character as a solvent is intermediate between a truly nonpolar solvent, such as hexane, and weakly polar solvents. Because the molecule is nonpolar, it is often classified as a nonpolar solvent, but it has some limited affinity with polar solutes because of its large molecular quadrupole. To improve its affinity with polar molecules further, carbon dioxide is sometimes modified with polar entrainers (*see* **Subheading 3.**). However, pure carbon dioxide can be used for many organic solute molecules even if they have some polar character. It has a particular affinity for fluorinated compounds and is useful for working with fluorinated metal complexes and fluoropolymers.

Carbon dioxide is not such a good solvent for hydrocarbon polymers and other hydrocarbons of high molar mass. Ethane, ethene, and propane become alternatives for these compounds, although they have the disadvantages of being hazardous because of flammability and of being somewhat less environmentally friendly. However, small residues of lower hydrocarbons in foodstuffs and pharmaceuticals are not generally considered a problem. Water has good environmental and other advantages, although its critical parameters are much less convenient (**Table 1**) and it gives rise to corrosion problems. Supercritical water is being used, at a research level, as a medium for the oxidative destruction of toxic waste *(2)*. There is a particular interest in both supercritical and near-critical water because of the behavior of its polarity. Ammonia has similar behavior, is often considered and discussed, but not often used. Many halocarbons have the disadvantage of cost or of being environmentally unfriendly. Xenon is expensive, but is useful for small-scale experiments involving spectroscopy because of its transparency in the infrared, for example *(3)*.

2. Properties of Supercritical Fluids

Although often pursued in practice for environmental reasons, the more fundamental interest in supercritical fluids arises because they can have properties intermediate between those of typical gases and liquids. Compared with liquids, densities and viscosities are less and diffusivities greater. Furthermore, properties are controllable by both pressure and temperature and the extra degree of freedom, compared with a liquid, can mean that more than one property can be optimized. Any advantage has to be weighed against the cost and inconvenience of the higher pressures needed. Consequently, supercritical fluids are exploited in particular areas.

A supercritical fluid changes from gaslike to liquidlike as the pressure is increased, and its thermodynamic properties change in the same way. Close to the critical temperature, this change occurs rapidly over a small pressure range. The most familiar property is the density, and its behavior is illustrated in **Fig. 2** *(4)*. This shows three density–pressure isotherms, and at the lowest temperature, 6 K above the critical temperature, the density change is seen to increase rapidly at around the critical pressure. As the temperature is raised, the change is less dramatic and moves to higher pressures. One consequence is that it is difficult to control the density near the critical temperature and, as many effects are correlated with the density, control of experiments and processes can be difficult. Other properties, such as enthalpy, also show these dramatic changes near the critical temperature.

The behavior of density, as well as all other thermodynamic functions, as a function of pressure and temperature can be predicted by an *equation of state*. Some of these have an analytical form, but the most accurate equations are complex numerical forms that have been obtained by intelligent fitting of a wide range of thermodynamic data, such as is carried out at the International

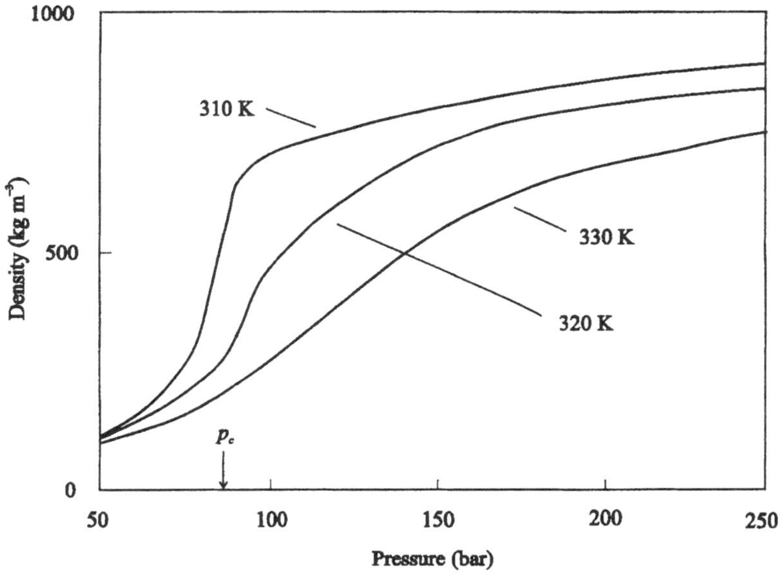

Fig. 2. Density–pressure isotherms for carbon dioxide *(4)*.

Union of Pure and Applied Chemistry Thermodynamic Tables Project Centre at Imperial College in London. They have carried out a study for a number of gases suitable as supercritical fluids. For carbon dioxide, a recent equation of state is that published by Span and Wagner *(5)*.

At low pressures, below 1 atm, the (dynamic) viscosity, η, of a gas is approximately constant, but thereafter rises with pressure in a similar way to density, ρ. However, the dependencies of density and viscosity on pressure at constant temperature are not conformal. A comprehensive correlation for the viscosity of carbon dioxide has been published *(6)*. **Table 2** shows typical values for the density and viscosity of a gas, supercritical fluid and liquid, taking carbon dioxide as an example. Using the example given, the viscosity of a supercritical fluid is much closer to that of a gas than that of a liquid. Thus, pressure drops across chromatographic columns and through supercritical extraction and other processes are less than for the equivalent liquid processes.

Diffusion coefficients, also shown in **Table 2** for naphthalene in carbon dioxide, are higher in a supercritical fluid than in a liquid. They are approximately inversely proportional to the fluid density *(7)*. The advantage shown in the table is seen not to be so great and the main diffusional advantage lies in the fact that typical supercritical solvents have lighter molecules than those of typical liquid solvents. The diffusion coefficient for naphthalene in a typical liquid would be about 1×10^{-9} m^2 s^{-1}. Thus diffusion coefficients in supercritical

Table 2
Typical Values of Density, Viscosity, and the Diffusion Coefficient Using Carbon Dioxide as Example

	CO_2		Naphthalene in CO_2
	Density *(4)* ρ (kg m^{-3})	Viscosity *(5)* η (μPa s)	Diffusion Coeff. *(6)* D (m^2 s^{-1})
Gas, 313 K, 1 bar	2	16	5.1×10^{-6}
Supercritical, 313 K, 100 bar	632	17	1.4×10^{-8}
Liquid, 300 K, 500 bar	1029	133	8.7×10^{-9}

fluid experiments and processes are typically an order of magnitude higher than in a liquid medium. This has advantages in band-narrowing in chromatography and faster transport in extraction. However, diffusion coefficients tend to zero at the critical point and fall in the critical region around it. At high concentrations, this can cause chromatographic band-broadening near the critical density *(8)*.

Although the S. I. unit of pressure is the pascal (Pa), it is rarely used in the field of supercritical fluids because of the high pressures involved. A more appropriate unit is the megapascal (MPa). Furthermore, no one pressure unit predominates; a wide variety are used interchangeably throughout the world. To help clear the confusion, the following may be of use: 1 atm = 1.0132 bar = 0.10132 MPa = 14.696 psi = 1.0332 kg/cm^2.

3. Modifiers

The solvent characteristics of a fluid can be modified by adding a modifier (also known as an entrainer or cosolvent) and this has been most commonly done with carbon dioxide. As this molecule is nonpolar, it is classified as a nonpolar solvent, although it has some limited affinity with polar solutes because of its large molecular quadrupole. Thus, pure carbon dioxide can be used for many large organic solute molecules, even if they have some polar character. For the extraction and chromatography of more polar molecules, it is common to add polar modifiers, such as the lower alcohols. Modifiers can also be added to develop other characteristics. They can impart increased or decreased polarity, aromaticity, chirality, and the ability to further complex metal-organic compounds. Just as carbon dioxide is the most popular substance for use as a supercritical fluid, it is also the substance to which modifiers are most frequently added. This is because modifiers are seen as a way of making use of this desirable compound in circumstances where it is not the best solvent. For example, in the case of carbon dioxide, methanol is added to increase polarity, aliphatic hydrocarbons to decrease it, toluene to impart aromaticity,

Table 3
Substances Useful as Modifiers in Carbon Dioxide with Critical Parameters

Substance	Critical Temperature, T_c (K)	Critical Pressure, p_c (bar)
Methanol	513	81
Ethanol	514	61
1-Propanol	537	51
2-Propanol	508	48
2-Butanol	536	42
Acetone	508	47
Acetonitrile	546	48
Acetic acid	593	58
Diethyl ether	467	36
Dichloromethane	510	63
Chloroform	536	54
Hexane	508	30
Benzene	562	49
Toluene	592	41
Tributyl phosphate	742	24

Data from **ref.** ***1***.

[*R*]-2-butanol to add chirality, and tributyl phosphate to enhance the solvation of metal complexes. They are often added in 5% or 10% amounts by volume, but sometimes much more, say 50%. They can have significant effects when added in small quantities, and in these cases it may be the effect on surface processes rather than solvent character, which is important. For example, the modifier may be effective in extraction by adsorbing on to surface sites, preventing the readsorption of a compound being extracted. Similarly, in chromatography, the modifier may cap active or unbonded sites on a stationary phase, preventing tailing of chromatographic peaks. A comprehensive review of modifiers has been made by Page et al. ***(9)***. Some compounds commonly used as modifiers are listed with their critical parameters in **Table 3**. It is important to be aware of the modifier-fluid phase diagram to ensure that the solvent is in one phase. For example, for methanol–carbon dioxide at 50°C there is only one phase above 95.5 bar whatever the composition, but below this pressure, two phases can occur. Above this pressure, the character of the medium depends on the proportions of modifier and fluid substance. If the proportion of modifier is low, the mixture will have the characteristics of a supercritical fluid, but if it is high, the medium will be liquidlike.

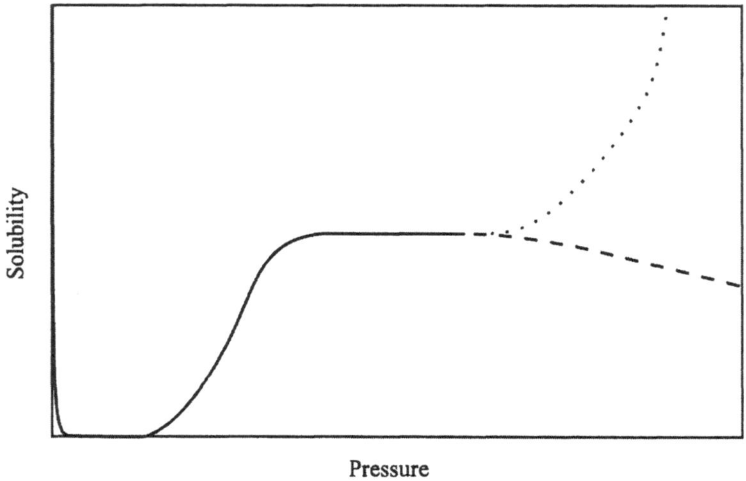

Fig. 3. The behavior of solubility in a supercritical fluid, shown schematically.

4. Solubility in a Supercritical Fluid

The behavior, at constant temperature, of the solubility of a substance in a supercritical fluid, in terms of mole fraction, is illustrated schematically in **Fig. 3**. When the pressure is close to zero, only the solute is present as vapor, and the mole fraction of the solute is unity. There is then an initial fall almost to zero at very low pressures as the solvent is added, and the solute is diluted without being much solvated. After staying close to zero, there is then a rise in solubility at around the critical density of the fluid, that is, when the density is increasing rapidly with pressure. This rise is due to solvation originating from attractive forces between the solvent and solute molecules. Thereafter, the solubility may exhibit a fall, represented by the dashed line. If this occurs, it is because at higher pressures, the system is becoming compressed and repulsive solute–solvent interactions are important. The solute can be said to be "squeezed out" of the solvent. Alternatively, a rise may occur, as represented by the dotted line. This happens if there is a critical line present at high pressures at the temperature of the isotherm and the solubility will rise toward it. The rising type of curve is a feature of smaller more volatile molecules and higher temperatures and vice versa. All situations between the two curves occur.

Correlation of supercritical fluid solubility data is not straightforward. All the features shown in **Fig. 3** can be reproduced qualitatively by any equation of state. For quantitative fitting, more refined equations of state are useful in certain regions, and, of these, the Peng-Robinson has been the most widely used. However, even this equation is not successful in fitting all the data at all pressures and temperatures. A further problem is that the parameters necessary for using the equation of state are not always available. Thus, often empirical approaches are used ***(10)***.

5. Applications of Supercritical Fluids

The areas where supercritical fluids are used are as follows and their advantages above the general ones of less pollution in the working and general environment and less solvent disposal costs are also given.

The most popular applications of supercritical fluids are as media for extraction (*see* Chapters 2–17, and 29) and chromatography (*see* Chapters 16–22).

5.1. Supercritical Fluid Extraction

Supercritical fluid extraction (SFE) uses a supercritical fluid to remove soluble substances from insoluble matrices. Supercritical fluids have attractive properties for extraction (*see* **Subheading 2.**) because they not only penetrate a sample faster than liquid solvents (supercritical fluids have diffusion coefficients midway between gases and liquids) and transport extracted material from the sample faster (supercritical fluids have viscosities like those of gases), but they also dissolve solutes from a sample matrix (supercritical fluids have solvating powers approaching those of liquids). Another advantage of SFE is less solvent residues in products.

The basic concept of SFE is to use a relatively cheap and safe material for the extraction of organic compounds from a matrix in place of conventional solvent extraction, cutting down on manipulation and avoiding the problems associated with the use and disposal of organic solvents. Although a number of substances are considered as potentially useful for SFE, in practice, the one of choice is carbon dioxide for the reasons given earlier (*see* **Subheading 1.**).

All designs of SFE apparatus, regardless of complexity and cost, share the same basic components: a source of extraction fluid, one or more pumps, a sample cell, an oven, a back-pressure regulator (BPR), and a collector (**Fig. 4**). The solvent delivery system consists of a pump to deliver liquid carbon dioxide and, optionally, a pump to supply modifier. The oven is used to keep the cell contents above the critical temperature of the extraction fluid. An equilibration coil is included to help mixing of carbon dioxide and modifier and aid thermal equilibration of the extraction fluid and the insides of the oven. The cell, usually a hollow stainless steel cylinder, is housed in the oven and contains the sample to be extracted. It has a frit at both ends to prevent insoluble material leaving the cell, but allowing soluble substances to pass through unhindered. The BPR serves to keep the pressure in the system above the critical pressure of the extraction fluid. It is, typically, a length of fused silica capillary (50 μm i.d.) or a mechanical or electronic needle valve. The silica restrictor is

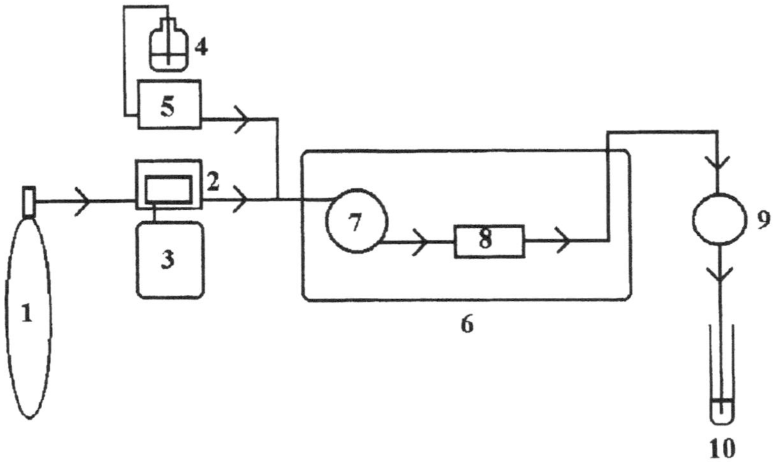

Fig. 4. Schematic diagram of a simple supercritical fluid extractor. 1, Source of carbon dioxide; 2, carbon dioxide pump; 3, chiller unit; 4, modifier reservoir; 5, modifier pump; 6, oven; 7, equilibration coil; 8, cell; 9, back-pressure regulator; and 10, collector.

usually connected by a graphite ferrule to a union attached to a length of 1/16-inch stainless steel tube coming from the sample cell. The BPR is heated (with a hairdryer or in an oven) to reduce the frequency of blockages by, for example, the formation of ice. Finally, a collection system is required to trap extracted material. It is usually a solid trap or a small glass collector containing a few cubic centimeters of organic solvent.

During an extraction, carbon dioxide and, optionally, modifier are pumped at set flow rates through a cell containing the sample. Soluble components of the sample are dissolved and removed from the cell. The extracted materials pass through the BPR and are depressurized into a collector containing a few cubic centimeters of organic solvent. The contents of the collector are evaporated to dryness or adjusted to a known volume, prior to analysis by, for example, supercritical fluid chromatography. An alternative way of collecting the extract is to depressurize it onto a packed trap. The solutes are then rinsed from the trap with an appropriate solvent into a small vial, ready for analysis or evaporation to dryness. This is known as off-line SFE. The extract can alternatively be fed directly into an analytical instrument in so-called on-line mode.

There are two different types of SFE: dynamic and static. In dynamic SFE, the supercritical fluid is pumped through the cell containing the sample continually. In the static mode, the sample is bathed in supercritical fluid, and there is no flow of fluid to or from the cell during the extraction.

Sometimes, both types of SFE are performed on the same sample at different times during an extraction *(11)*.

Extraction with supercritical fluids has been applied to many matrices, such as fossil fuels *(12,13)*, environmental pollutants *(14,15)*, natural products *(16)*, foods *(17)*, drugs *(18)*, metals *(19)* and polymers *(20,21)*. Extraction was the first commercial use of supercritical fluids, and examples include the extraction of hops *(4)* and the decaffeination of coffee (*see* Chapter 2). More than 400 research papers have been produced on the extraction of a wide range of natural products, including high-value pharmaceutical precursors *(22)*. Fractionation of liquid mixtures can be achieved by countercurrent extraction, and this can be improved by imposing a temperature gradient on the column, which causes refluxing to occur *(23)*. It is largely applied to natural products, such as essential oils and lipid products, and can be used to concentrate substances before chromatography. The advantage of using a supercritical fluid is that countercurrent extraction with reflux can be carried out in one unit. The most successful applications of SFE have been for relatively nonpolar compounds. Some polar compounds have presented problems *(24)*, but efforts have been made to make SFE viable *(25)*.

5.2. Supercritical Fluid Chromatography

Supercritical fluid chromatography (SFC) can be defined as the separation of organic compounds using a supercritical fluid as the mobile phase. There is interest in the technique because the rapid diffusivity and low viscosity of a supercritical fluid allows faster separations and better resolution of components in a solution than high performance liquid chromatography (HPLC). Furthermore, sensitive general detectors, like the flame ionization detector (FID), can be exploited. Chromatography with supercritical fluids can be an ecofriendly alternative to HPLC, which uses moderate volumes of toxic organic solvents, and a more versatile substitute for gas chromatography, which is limited to volatile organic compounds. Another advantage of SFC can be little or no solvent residues in products.

Not surprisingly, carbon dioxide is the most common mobile phase in SFC. Its low critical temperature allows the separation of thermally sensitive compounds, but supercritical carbon dioxide is not very polar, limiting its use as a solvent. To overcome this, carbon dioxide can be modified with polar organic solvents such as methanol, but this tactic renders the FID redundant.

Chromatography with supercritical fluids has been used with packed and capillary columns. Compatibility with the FID means that SFC can be used for samples that would be difficult to detect by HPLC. The technique is relatively easy to couple to other instruments, for example, a Fourier transform infrared spectrometer *(26)*. Chromatography with supercritical fluids has been performed on an analytical *(27)* and a preparative scale *(28)*.

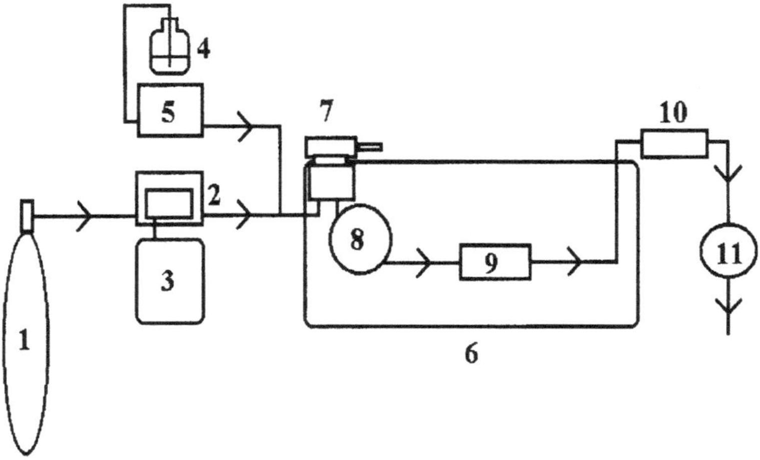

Fig. 5. Basic instrument used for supercritical fluid chromatography. 1, Source of carbon dioxide; 2, carbon dioxide pump; 3, chiller unit; 4, modifier reservoir; 5, modifier pump; 6, oven; 7, injection valve; 8, equilibration coil; 9, column; 10, detector; and 11, back-pressure regulator.

The instrumentation used for SFC is similar to SFE apparatus (*see* **Subheading 5.1.**), but there are differences. A column containing stationary phase replaces the sample-holding cell. Furthermore, SFC systems include an injector just before the column and a detector between the column and the back-pressure regulator (**Fig. 5**). The mobile phase is initially pumped as a liquid until it reaches the oven, where it becomes a supercritical fluid. The oven houses the body of the injector and the column, and keeps them above the critical temperature of the substance used as the mobile phase. The sample in liquid solvent is injected into the mobile phase and passes on to the column where its constituents are separated. From the column, the isolated components pass into the detector (still under considerable pressure) before entering the back-pressure regulator and on to waste or collection. Here, the depressurized fluid becomes a gas (carbon dioxide) and, if modifier is used, a liquid.

A wide range of compounds have been separated and/or analyzed by SFC. Examples include cholesterol ***(29)***, polymer additives ***(21)*** and oligomers ***(30)***, bile acids ***(31,32)***, ecdysteroids ***(33)***, azadirachtin ***(34)***, acidic drugs ***(35)***, and basic drugs ***(35)***. Chromatography with supercritical fluids can be applied to chiral separations (*see* Chapter 19) and high-value products (*see* Chapters 20–22). Efficient simulated bed units are available ***(36)***. However, SFE and SFC are not the only uses for supercritical fluids.

5.3. Miscellaneous Applications

Chemical reactions in supercritical fluids (*see* Chapters 23–25) are being researched with some in production *(37)*. There is interest in this area because supercritical fluids can homogenize a reaction mixture, diffusion is more rapid for diffusion-controlled reactions, they can incorporate controlled phase separation of products, and, especially in the critical region, they can be used to control the distribution of products. Metals processing, using complexing agents in the supercritical fluid, is also being researched *(38)*. Supercritical fluids can be used in environmental clean-up methods, including soil remediation *(39)*, by removal of both organics and metals, and effluent treatment by supercritical water oxidation *(40)*. Painting and coating, with carbon dioxide as part-solvent, is used in production *(41)*. Impregnation and dyeing of polymers and synthetic fibers with supercritical fluids is established and the dyeing of cotton is being researched, with the advantage of considerable reduction in water pollution *(42)*. The use of supercritical fluids for particle formation in the micrometer range with a narrow size distribution can be carried out (*see* Chapters 26–28). The advantage of this method is the absence of degradation by heating during the alternative milling process. Cleaning of high-value electrical and mechanical components can be carried out with supercritical fluids *(43,44)*. Another advantage of supercritical fluids is the absence of surface tension, improving penetration and avoiding distortion of delicate components during drying (*see* Chapter 31).

This chapter gives only a brief introduction to supercritical fluids. Much more comprehensive texts are available, for example, those by McHugh and Krukonis *(4)* and Smith *(45)*.

References

1. Reid, R. C., Prausnitz, J. M., and Poling, B. E. (1986) *The Properties of Gases and Liquids*. McGraw-Hill, New York.
2. Modell, M. (1982) Processing methods for the oxidation of organics in supercritical water. *U.S. Patent* 4,338,199.
3. Howdle, S. M., Healy, M. A., and Poliakoff, M. (1990) Organometallic chemistry in supercritical fluids: the generation and detection of dinitrogen and non-classical dihydrogen complexes of group 6, 7 and 8 transition metals at room temperature. *J. Am. Chem. Soc.* **112,** 4804–4813.
4. McHugh, M. A. and Krukonis, V. J. (1994) *Supercritical Fluid Extraction*, 2nd ed., Butterworth-Heinemann, Boston.
5. Span, R. and Wagner, W. (1996) A new equation of state for carbon dioxide covering the fluid region from the triple-point temperature to 1100 K at pressures up to 800 MPa. *J. Phys. Chem. Ref. Data* **25,** 1509–1596.
6. Vesovic, V., Wakeham, W. A., Olchowy, G. A., Sengers, J. V., Watson, J. T. R., and Millat, J. (1990) The transport properties of carbon dioxide. *J. Phys. Chem. Ref. Data* **19,** 763–808.

7. Clifford, A. A. and Coleby, S. E. (1991) Diffusion of a solute in dilute solution in a supercritical fluid. *Proc. R. Soc. Lond.* A**433,** 63–79.
8. Bartle, K. D., Baulch, D. L., Clifford, A. A., and Coleby, S. E. (1991) Magnitude of the diffusion coefficient anomaly in the critical region and its effect on supercritical fluid chromatography. *J. Chromatogr.* **557,** 69–83.
9. Page, S. H., Sumpter, S. R., and Lee, M. L. (1992) Fluid phase equilibria in supercritical fluid chromatography with CO_2-based mixed mobile phases: a review. *J. Microcol. Sep.* **4,** 91–122.
10. Bartle, K. D., Clifford, A. A., Jafar, S. A., and Shilstone, G. F. (1991) Solubilities of solids and liquids of low volatility in supercritical carbon dioxide. *J. Phys. Chem. Ref. Data* **20,** 713–756.
11. Heikes, D. L. (1994) SFE with GC and MS determination of safrole and related allylbenzenes in sassafras teas. *J. Chromatogr. Sci.* **32,** 253–258.
12. Supercritical Fluid Technology Synopsis, SFE-87, Suprex Corporation, Pittsburgh, PA, 1991.
13. Isco Applications Bulletin 71, Isco Inc., Lincoln, Nebraska, 1991.
14. Janda, V., Bartle, K. D., and Clifford, A. A. (1993) Supercritical fluid extraction in environmental analysis. *J. Chromatog. A* **642,** 283–299.
15. Barnabas, I. J., Dean, J. R., and Owen, S. P. (1994) Supercritical fluid extraction of analytes from environmental samples: a review. *Analyst* **119,** 2381–2394.
16. Smith, R. M. (1996) Supercritical fluid extraction of natural products. *LC-GC Intl.* **9,** 8–15.
17. Um, K. W., Bailey, M. E., Clarke, A. D., and Chao, R. R. (1992) Concentration and identification of volatile compounds from heated beef fat using supercritical CO_2 extraction-gas liquid chromatography/mass spectrometry. *J. Agric. Food Chem.* **40,** 1641–1646.
18. Cirimele, V., Kintz, P., Majdalani, R., and Mangin, P. (1995) Supercritical fluid extraction of drugs in drug addict hair. *J. Chromatog. B* **673,** 173–181.
19. Lin, Y. and Wai, C. M. (1994) Supercritical fluid extraction of lanthanides with fluorinated β-diketones and tributyl phosphate. *Anal. Chem.* **66,** 1971–1975.
20. Via, J. C., Braue, C. L., and Taylor, L. T. (1994) Supercritical fluid fractionation of a low molecular weight, high-density polyethylene wax using carbon dioxide, propane, and propane-modified carbon dioxide. *Anal. Chem.* **66,** 603–609.
21. Hunt, T. P., Dowle, C. J., and Greenway, G. (1991) Analysis of poly(vinyl chloride) additives by supercritical fluid extraction and supercritical fluid chromatography. *Analyst* **116,** 1299–1304.
22. Sangün, M. K. (1998) Selective supercritical fluid extraction from plant materials. Ph.D. thesis. School of Chemistry, Leeds University, UK.
23. Sato, M., Goto, M., Kodama, A., and Hirose, T. (1997) Supercritical fluid extraction with reflux for citrus oil processing. *ACS Symp. Ser.* **670,** 119–131.
24. Cross, R. F., Ezzell, J. L., and Richter, B. E. (1993) The supercritical fluid extraction of polar drugs (sulfonamides) from inert matrices and meat animal products. *J. Chromatogr. Sci.* **31,** 162–169.
25. Luque de Castro, M. D. and Tena, M. T. (1996) Strategies for supercritical fluid extraction of polar and ionic compounds. *Trends Anal. Chem.* **15,** 32–37.

26. Ashraf-Khorassani, M., Combs, M. T., Taylor, L. T., Willis, J., Liu, X. J., and Frey, C. R. (1997) Separation and identification of sulfonamide drugs via SFC/FT-IR mobile phase elimination interface. *App. Spectros.* **51,** 1791–1795.
27. Taylor, L. T. (1997) Trends in supercritical fluid chromatography: 1997. *J. Chromatogr. Sci.* **35,** 374–382.
28. Bartle, K. D., Bevan, C. D., Clifford, A. A., Jafar, S. A., Malak, N., and Verrall, M. S. (1995) Preparative-scale supercritical fluid chromatography. *J. Chromatogr. A* **697,** 579–585.
29. Nomura, A., Yamada, J., Takatsu, A., Horimoto, Y., and Yarita, T. (1993) Supercritical fluid chromatographic determination of cholesterol and cholesteryl esters in serum on ODS-silica gel column. *Anal. Chem.* **65,** 1994–1997.
30. Bartle, K. D., Boddington, T., Clifford, A. A., and Cotton, N. J. (1991) Supercritical fluid extraction and chromatography for the determination of oligomers in poly(ethylene terephthalate) films. *Anal. Chem.* **63,** 2371–2377.
31. Scalia, S. and Games, D. E. (1993) Determination of free bile acids in pharmaceutical preparations by packed column supercritical fluid chromatography. *J. Pharm. Sci.* **82,** 44–47.
32. Villette, V., Herbreteau, B., Lafosse, M., and Dreux, M. (1996) Free bile acid analysis by supercritical fluid chromatography and evaporative light scattering detection. *J. Liq. Chrom. Rel. Technol.* **19,** 1805–1818.
33. Morgan, E. D., Murphy, S. J., Games, D. E., and Mylchreest, I. C. (1988) Analysis of ecdysteroids by supercritical fluid chromatography. *J. Chromatogr.* **441,** 165–169.
34. Huang, H. P. and Morgan, E. D. (1990) Analysis of azadirachtin by supercritical fluid chromatography. *J. Chromatogr.* **519,** 137–143.
35. Roberts, D. W., Wilson, I. D., and Reid, E. (1990) *Methodol. Surv. Biochem. Anal.* **20,** 257.
36. Mazzotti, M., Storti, G., and Morbidelli, M. (1997) Supercritical fluid simulated moving bed chromatography. *J. Chromatogr. A* **786,** 309–320.
37. Fukuzato, R. (1991) Supercritical fluid processing research and business activities in Japan In *Proceedings of the second international symposium on supercritical fluids* (McHugh, M. A., ed.), John Hopkins University Press, Baltimore, p. 196.
38. Wai, C. M. and Wang, S. F. (1997) Supercritical fluid extraction: metals as complexes. *J. Chromatogr. A* **785,** 369–383.
39. Ekhtera, M. R., Mansoori, G. A., Mensinger, M. C., Rehmat, A., and Deville, B. (1997) Supercritical fluid extraction for remediation of contaminated soil. *ACS Symp. Ser.* **670,** 208–231.
40. Mitton, D. B., Han, E. H., Zhang, S. H., Hautanen, K. E., and Latanisian, R. M. (1997) Degradation in supercritical water oxidation systems. *ACS Symp. Ser.* **670,** 242–254.
41. Donohue, M. D., Geiger, J. L., Kiamos, A. A., and Nielsen, K. A. (1996) Reduction of volatile organic compound emissions during spray painting: a new process using supercritical carbon dioxide to replace traditional paint solvents. *ACS Symp. Ser.* **626,** 152–167.
42. Özcan, A. S., Clifford, A. A., and Bartle, K. D. (1998) Dyeing of cotton fibres with disperse dyes in supercritical carbon dioxide. *Dyes Pigments* **36,** 103–110.

43. Bakker, G. L. and Hess, D. W. (1998) Surface cleaning and carbonaceous film removal using high pressure, high temperature water and water/CO_2 mixtures. *J. Electrochem. Soc.* **145,** 284–291.
44. Cooney, C. M. (1997) Supercritical CO_2-based cleaning system among Green Chemistry Award winners. *Environ. Sci. Tech.* **31,** A314–A315.
45. Smith, R. M., ed. (1988) *Supercritical Fluid Chromatography.* The Royal Society of Chemistry, London.

2

Supercritical Fluid Extraction of Caffeine from Instant Coffee

John R. Dean, Ben Liu, and Edwin Ludkin

1. Introduction

Caffeine—1,3,7-trimethylxanthine—is one of three common alkaloids found in coffee, cola nuts, tea, cacao beans, maté, and other plants. The other two are theophylline and theobromine *(1)*. The effects of caffeine are commonly reported to be as a stimulant of the central nervous system, cardiac muscle, and the respiratory system. It is also a common diuretic and delays fatigue *(1)*. It has also been reported *(1)* that caffeine in combination with an analgesic, for example, aspirin, can be used in the treatment of headaches. However, there are few data to substantiate its efficacy in this role.

The concept of supercritical fluid extraction (SFE) was introduced in Chapter 1. Extraction with supercritical carbon dioxide (CO_2) as the solvent has been used to isolate components from different matrices such as biological and environmental samples *(2)*. The commercial process of extraction of caffeine from coffee using supercritical CO_2 was patented by Zosel in 1964 *(3)*. The analytical SFE of caffeine from coffee has been reported by other workers using SFE coupled to supercritical fluid chromatography *(4)*, nuclear magnetic resonance spectroscopy *(5)*, infrared spectroscopy *(6)*, and high performance liquid chromatography (HPLC) *(7)*. However, the use of a nonpolar supercritical fluid, such as CO_2, does not exhaustively extract caffeine from instant coffee. As has been reported elsewhere *(2)*, the polarity of the supercritical fluid can be increased by the addition of a polar organic solvent, for example, methanol. This approach is commonly used for "real" sample analysis.

The purpose of this chapter is to describe a procedure for the off-line SFE of caffeine from instant coffee granules using supercritical CO_2-methanol and to provide an introductory practical/training exercise in the application of SFE. Analysis of the extracts is done by HPLC with ultraviolet detection.

From: *Methods in Biotechnology, Vol. 13: Supercritical Fluid Methods and Protocols*
Edited by: J. R. Williams and A. A. Clifford © Humana Press Inc., Totowa, NJ

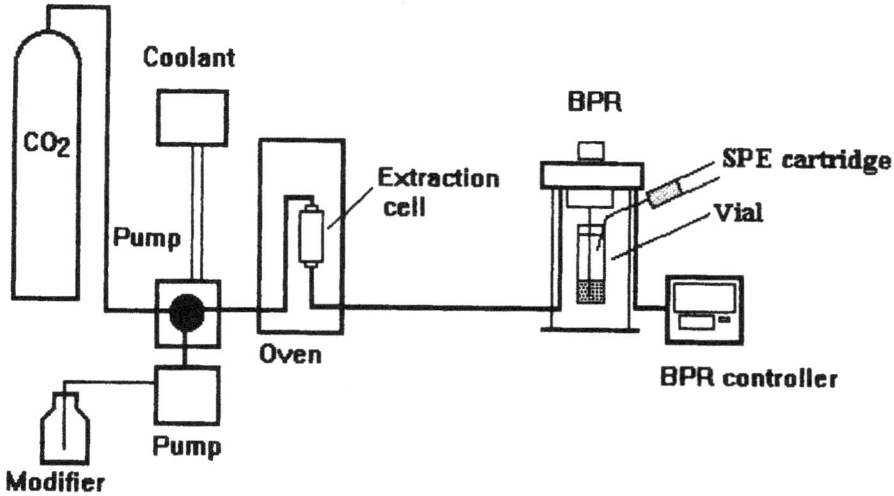

Fig. 1. Schematic diagram of the SFE apparatus.

2. Materials

2.1. SFE

1. Two reciprocating pumps (*see* **Fig. 1**); one to deliver CO_2 and the other to dispense modifier (intelligent HPLC pumps, model PU-980, Jasco Ltd., Great Dunmow, Essex, U.K.).
2. A column oven (Jasco, model 860-CO) which can operate up to 100°C (*see* **Fig. 1**).
3. A back-pressure regulator (see **Fig. 1**) or BPR (Jasco, model 880–81).
4. A recirculating water bath containing an ethylene glycol mixture, which is passed through a jacket that encases the CO_2 pump-head only (*see* **Fig. 1**).
5. An extraction cell (*see* **Fig. 1**).
6. Analyte collection occurs during depressurization into a glass collection vial containing a suitable organic solvent (methanol) fitted with a rubber septum through which two holes are pierced (*see* **Fig. 1**). Into one hole passes the connecting tube from the BPR, while the other contains a syringe needle fitted with a solid-phase extraction (SPE) cartridge (C18, Waters Sep-Pak, Millipore Co., Milford, MA). The purpose of the latter is to prevent loss of analyte from the collection vial and to vent the escaping gaseous carbon dioxide.
7. SFE-grade CO_2, fitted with a diptube (Air Products Ltd., Sunderland, UK).
8. HPLC-grade methanol.
9. Celite (Celite for GLC, Merck Ltd., Poole, Dorset, U.K.).

2.2. HPLC

1. Reciprocating pump (Gilson, model 305, Anachem Ltd., Luton, Beds, UK).
2. Separation column (C18, ODS2, 25 cm × 4.6 mm, Phase Separations Ltd., Clwyd) maintained at a temperature of 35°C.

3. Injection volume, 20 µL.
4. Mobile phase, acetonitrile:water:acetic acid (15:84:1), was pumped at a flow rate of 1 mL/min.
5. An ultraviolet-visible detector (Jasco, UV-975) for monitoring the response at a wavelength of 275 nm.

2.3. Sample

1. Instant coffee granules were purchased from local retail outlets in both decaffeinated and caffeinated forms.

3. Method

3.1. Sample Preparation

1. Grind instant coffee granules into powder using a mortar and pestle, and sieve through a 420-µm filter.
2. Mix one part of the ground instant coffee with one part of Celite (*see* **Note 1**).

3.2. SFE

1. Turn on the electrical supply to the SFE system, including the recirculating water bath. Allow 30 min for cooling of the CO_2 pump-head.
2. Take an extraction cell (*see* **Note 2**) and tighten, using a wrench, an end-cap on one end only and then weigh the cell.
3. Fill the extraction cell with the coffee/Celite mixture (approx 0.5–0.7 g), and weigh the cell again.
4. Tighten the other end-cap on to the cell with the wrench and insert the capped cell into the oven. Plumb the cell into the SFE system. This requires the use of a wrench to ensure a suitable connection.
5. Connect a glass collection vial containing 2 mL of methanol and fitted with a C18 SPE cartridge to the outlet of the BPR (*see* **Subheading 2.1., step 6**).
6. Set SFE operating parameters: flow rate of liquid carbon dioxide, 1.8 mL/min and methanol, 0.2 mL/min; oven temperature, 60°C; and pressure, 250 kg/cm^2. Allow the system to operate for a few minutes to establish a working system. Before the extraction commences, preheat the extraction cell containing the sample to the preset temperature for 10 min (*see* **Note 3**), then undergo a static extraction (no flow of CO_2) at the operating conditions for 5 min and, finally, a dynamic extraction (flow of CO_2 and methanol) for 1 h.
7. After the allotted extraction time, remove the collection vial from the system and back-flush the C18 SPE cartridge with 2 mL of fresh methanol (*see* **Note 4**).
8. Extract further samples using the stated parameters.

3.3. Analysis of Coffee for Caffeine

1. Quantitatively transfer the contents of the collection vial into a 25-mL volumetric flask and adjust to the required volume with a 1:1 water:methanol mixture (for decaffeinated products only). For caffeinated products, pipet 1 mL of

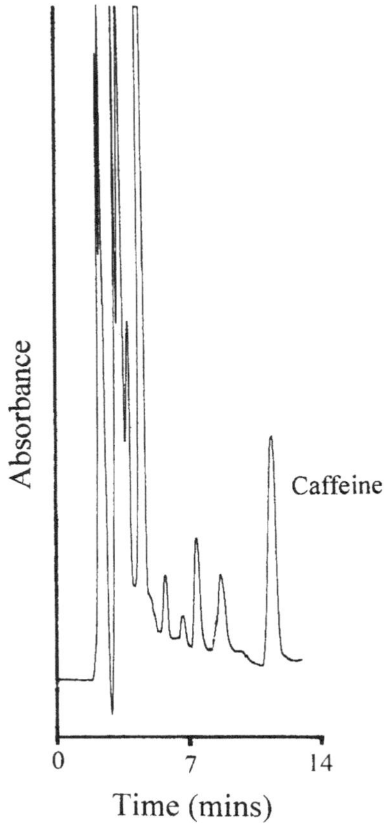

Fig. 2. HPLC chromatogram of caffeine extracted from decaffeinated instant coffee.

the diluted sample solution into another 25-mL volumetric flask and adjust to the required volume with water.

2. Analyze for caffeine using HPLC (*see* **Subheading 2.2.**) by first establishing a calibration graph for caffeine. This entails running a series of 4 to 5 caffeine standards of known concentration in methanol. There should be a linear relationship between absorbance and caffeine concentration over the concentration range of interest. The caffeine peak appears at a retention time of approximately 11 min.
3. Analyze for the unknown levels of caffeine in the coffee extracts.
4. Typical caffeine levels in commercial instant coffees (using four varieties for which decaffeinated and caffeinated were available and a single variety for which only decaffeinated was available) determined by off-line SFE–HPLC ranged from 0.131 ± 0.006% (w/w) to 0.058 ± 0.001% (w/w) for decaffeinated coffee and from 2.373 ± 0.115% (w/w) to 1.811 ± 0.241% (w/w) for caffeinated coffee (*see* **Note 5**). Typical chromatograms obtained for decaffeinated and caffeinated coffee extracts are shown in **Figs. 2** and **3**, respectively.

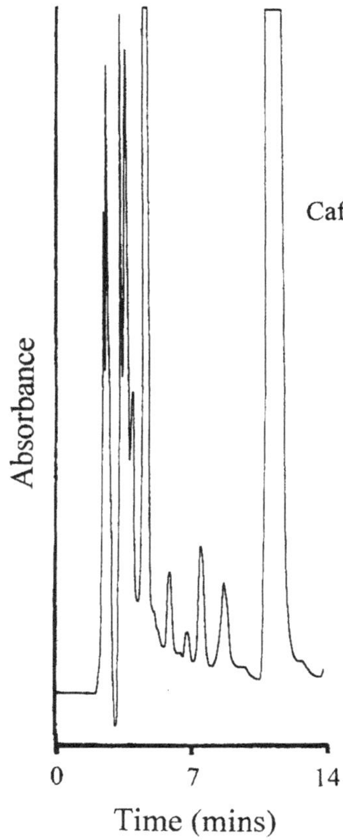

Fig. 3. HPLC chromatogram of caffeine extracted from caffeinated instant coffee.

4. Notes

1. Before the ground instant coffee is extracted using 10% methanol-modified supercritical CO_2, it should be dispersed with Celite. The grinding and mixing of the coffee with Celite serves to produce a larger surface area for solute–solvent interaction that is, caffeine-CO_2/methanol interaction.
2. Ensure the extraction cell is suitable for its purpose, that is, able to withstand high pressure and does not leak.
3. After insertion of the extraction cell into the oven, allow sufficient time for the cell and its contents to reach the preset temperature. Ten minutes was considered to be suitable in this experiment.
4. Back-flush the C18 SPE cartridge with 2 mL methanol after each extraction. This will ensure that quantitative analyses are performed.
5. Under the SFE conditions: pressure, 250 kg/cm^2; temperature, 60°C; extraction fluid, 10% methanol-modified CO_2; and a flow rate of 2 mL/min, it was possible

to extract approx 83% of caffeine from the ground instant coffee within 1 h, 89% in 2 h and 94% within 3 h (based on the recovery obtained after 5 h).

References

1. Lopez-Ortiz, A. (1997) Frequently asked questions about coffee and caffeine. internet address: http://www.cs.unb.ca/~alopez-o/caffaq.html
2. Dean, J. R. (1993) Applications of supercritical fluids in industrial analysis. Blackie Academic and Professional, Glasgow, U.K.
3. Zosel, K. (1964) German Patent 1,493,190.
4. Patrick, E., Masanori, Y., Yoshio, Y., and Maneo, S. (1991) Infrared and nuclear magnetic resonance spectrometry of caffeine in roasted coffee beans after separation by preparative supercritical fluid chromatography. *Anal. Sci.* **7,** 427–431.
5. Braumann, U., Handel, H., Albert, K., Ecker, R., and Spraul, M. (1995) On-line monitoring of the supercritical fluid extraction process with proton nuclear magnetic resonance spectroscopy. *Anal. Chem.* **67,** 930–935.
6. Heglund, D. L., Tilotta, D. C., Hawthorne, S. B., and Miller, D. J. (1994) Simple fiber-optic interface for on-line supercritical fluid extraction-Fourier transform infrared spectrometry. *Anal. Chem.* **66,** 3543–3551.
7. Ndiomu, C. F. and Simpson, C. F. (1988) Some applications of supercritical fluid extraction. *Anal. Chim. Acta* **213,** 237–243.

3

Supercritical Fluid Extraction of Nitrosamines from Cured Meats

John W. Pensabene and Walter Fiddler

1. Introduction

Supercritical fluid extraction or SFE (*see* Chapter 1) is used to isolate pesticides from environmental samples, fruits and vegetables. However, the use of this technique for the extraction of residues, such as nitrosamines at the ppb level, in cured meat products is relatively recent.

Of the 300 or more *N*-nitroso compounds tested, over 90% have been found to be carcinogenic *(1)*. The fact that nitrosamines induce cancer in at least 40 different animal species, including primates *(2)*, makes it likely that these compounds would also be active in humans. This accounts for the regulatory concern, the monitoring of, and establishment of tolerance or action levels for nitrosamine-containing foods.

The two SFE methods described in this chapter are alternatives to distillation *(3–5)* and solid-phase extraction or SPE *(6)* methods currently in use that employ considerable amounts of organic solvents, principally halogen-containing ones. Unlike the distillation methods, without the addition of a nitrosation inhibitor, SFE is not as susceptible to artifactual nitrosamine formation. These SFE methods for isolating volatile nitrosamines include *N*-nitrosopyrrolidine formed in bacon as a result of frying *(7)*, *N*-nitrosodibutylamine *(8)* and the semivolatile, *N*-nitrosodibenzylamine *(9–11)*, which is found primarily on the surface of boneless hams that are wrapped with rubber-containing elastic nettings. These methods are applicable to a wide range of cured meat products, from high fat bacon to lean boneless ham. For these three nitrosamines, and for the other

Mention of brand or firm names does not constitute an endorsement by the U.S. Department of Agriculture over others of a similar nature not mentioned.

From: *Methods in Biotechnology, Vol. 13: Supercritical Fluid Methods and Protocols*
Edited by: J. R. Williams and A. A. Clifford © Humana Press Inc., Totowa, NJ

N-nitroso compounds extracted, the methods use a chemiluminescence detector, a thermal energy analyzer [TEA *(12)*]. Interfacing this TEA to a gas chromatograph allows the separation and permits specific detection of *N*-nitroso compounds at the subnanogram level. SFE with off-line trapping on a commercial SPE cartridge is employed for the isolation of nitrosamines in both types of products. With a slight modification, the method for fried bacon is also applicable to its drippings. The procedure presented herein is simple, rapid, solvent-sparing, and offers a reproducible means for extracting nitrosamines from these complex food matrices.

2. Materials

1. The supercritical fluid extractor (SPE-ed SFE, Applied Separations, Allentown, PA, USA) was configured for the parallel extraction of two SFE vessels *(13)*. The pump was fitted with a recirculating chiller assembly (–10°C), for cooling the SFE pump-head, eliminating the need for helium-pressured carbon dioxide (CO_2) cylinders. Extraction vessels were connected to the system with hand-tightened, slip-free connectors (Keystone Scientific, Bellefonte, PA). Two 6 mL SPE cartridges (Applied Separations) containing 1.0 g of silica gel (*see* **Note 1**) were attached directly to the micrometering valves for off-line collection of the nitrosamines. A diagram of this instrument is shown in **Fig. 1**.
2. Supercritical-grade CO_2, without helium headspace.
3. High pressure (10,000 psi) extraction vessels, 24 mL capacity (Keystone Scientific).
4. Hydromatrix (Celite 566, *see* **Note 2**), propyl gallate, silica gel (*see* **Note 1**), dichloromethane (DCM), anhydrous diethyl ether, pentane, hexane (HPLC-grade).
5. Polypropylene wool (Aldrich Chemical Co, Milwaukee, WI).
6. Tamping rod and polyethylene frits for 24-mL extraction vessels (Applied Separations).
7. Floline SEF-51 flow meter-gas totalizer (Horriba, Sunnyvale, CA).
8. Concentrator tube (10 mL) and micro-Snyder columns (Kontes Glass Co, Vineland, NJ).
9. *N*-Nitrosodipropylamine (NDPA, *see* **Note 3**) internal standard solution, 0.10 μg/mL in DCM.
10. *N*-Nitrosodimethylamine (NDMA), N-nitrosodiethylamine (NDEA), NDPA, N-nitrosodibutylamine (NDBA), N-nitrosopiperidine (NPIP), N-nitrosopyrrolidine (NPYR), N-nitrosomorpholine (NMOR), each 0.10 μg/mL in DCM for bacon analysis (*see* **Note 4**).
11. NDPA, NDBA, *N*-nitrosodibenzylamine (NDBzA), each 0.10 μg/mL in DCM for ham analysis.
12. Quantitation method for bacon: Shimadzu gas chromatograph (GC) Model GC-14A equipped with a AOC-14 autoinjector or equivalent, and interfaced to a thermal energy analyzer (TEA) Model 502A chemiluminescence detector (Thermedics, Inc., Woburn, MA). The column used was a 2.7 m × 2.6 mm glass column packed with 15% Carbowax 20 M-TPA on 60–80 mesh Gas Chrom P. GC operating conditions: helium carrier gas, 35 mL/min; column program, 120°C to 220°C at

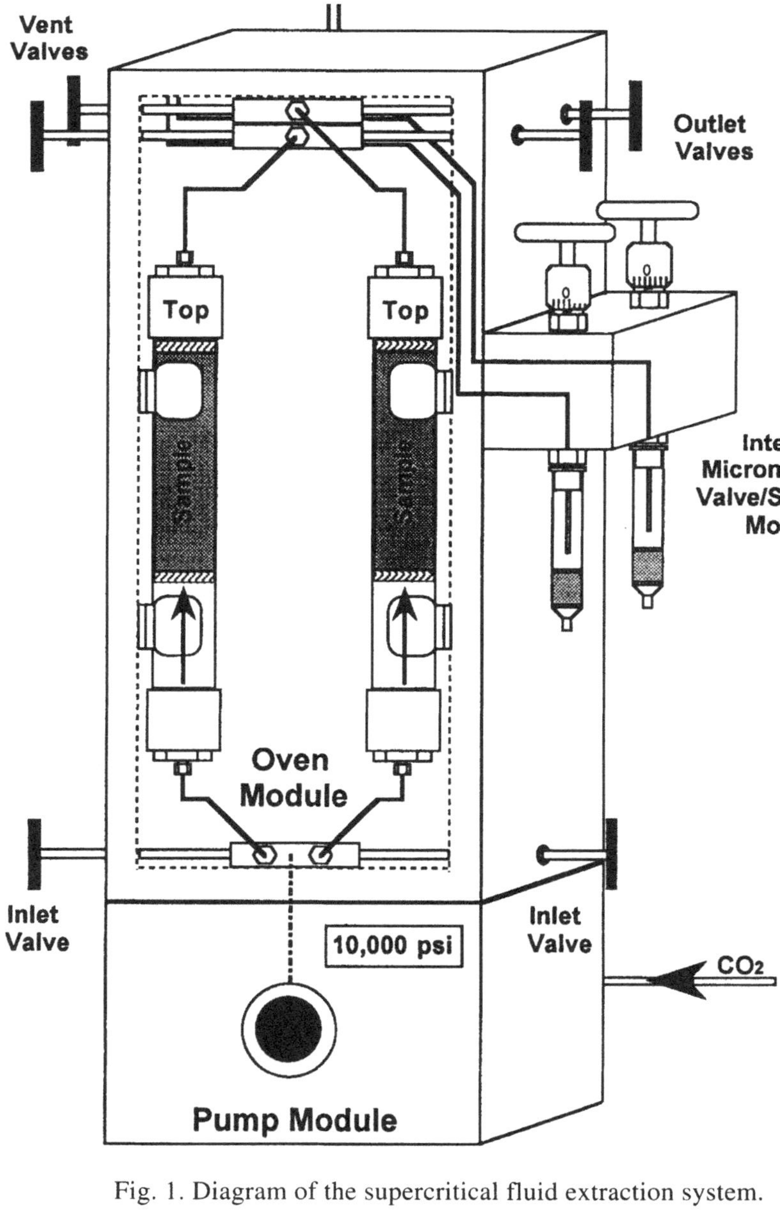

Fig. 1. Diagram of the supercritical fluid extraction system.

4°C/min; injector, 220°C. TEA conditions: furnace, 475°C; TEA vacuum, 1.0 m of mercury; liquid nitrogen cold trap.

13. Quantitation method for ham: Shimadzu Model GC-14A connected to an external pyrolyzer interface controlled by a TEA Model 610R Nitrogen Converter,

which in turn is interfaced to a TEA Model 502A. The column used was a 1.8 m × 2.6 mm glass column packed with 5% SP-2401 DB on 100–120 mesh Supelcoport. GC operating conditions: helium carrier gas, 35 mL/min; column program, 80°C for 3 min, then 10°C/min to 230°C; injector, 240°C. TEA conditions: pyrolyzer, 475°C; interface, 275°C; TEA vacuum, 0.8 mm of mercury; liquid nitrogen cold trap.

3. Method

1. Comminute and then mix the meat sample thoroughly to obtain a representative sample. All samples are to be analyzed in duplicate.
2. Weigh 5.0 g of meat sample ***(14,15)*** into a 100-mL beaker. Add 250 mg of propyl gallate to the sample to prevent artifactual nitrosamine formation.
3. Fortify the sample with 0.5 mL of NDPA internal standard using a 0.5 mL transfer pipette.
4. Add 5.0 g of Hydromatrix and stir mixture with a glass rod until it becomes a dry, free-flowing mixture (ca. 1 min).
5. Seal one end of the high-pressure extraction vessel and label it on top.
6. Add the dry, free-flowing sample mixture to the extraction vessel prepacked with a plug of polypropylene wool (*see* **Note 5**). Tightly compress the mixture with a tamping rod to ensure uniform supercritical fluid flow. Add a second plug of polypropylene wool to the extraction vessel and compress in place with the tamping rod (*see* **Note 6**). Seal bottom end of extraction vessel.
7. Install the extraction vessels in the SFE oven with the end labeled top connected to the upper fittings (**Fig. 1**).
8. Attach 6 mL SPE cartridges containing 1.0 g of silica gel to the micrometering valves (*see* **Note 7**). Attach the flow meter–gas totalizer to the SPE cartridges with flexible tubing. Ensure there are no leaks of gas at the connections.
9. Preheat the micrometering valves to 115°C. Close the outlet and vent valves; open the inlet valves.
10. Slowly pressurize the SFE vessels with CO_2 to approximately 8500 psi.
11. Set the oven temperature to 40°C (*see* **Note 8**), and equilibrate the system by using a 10-min static holding period.
12. Adjust the pressure to a final setting of 10,000 psi (680 bar).
13. After the 10-min heating period, open the outlet valves to direct flow through the micrometering valve module to the SPE cartridges. Use the micrometering valves to establish and maintain a 2.8 L CO_2/min (expanded gas) flow through the SPE cartridges during the extraction procedure.
14. After 50 L per vessel are recorded on the gas totalizer, close the inlet and outlet valves and depressurize the SFE vessels by slowly opening the vent valves.
15. Remove the extraction vessels from the oven, and attach Luer adapters to the upper slip-free connectors of the extractor. Attach a 1-mL glass syringe to each adapter and flush any trace residues of analyte-lipid remaining in the lines with 0.3 mL of hexane.

16. Remove the SPE cartridges containing the analyte–lipid mixture from the micrometering valve collection assembly. Hold the cartridges below the micrometering valve and rinse the external system with 0.1 mL of hexane directly into the SPE cartridges to ensure quantitative recovery of the nitrosamines.
17. Remove lipid material from the silica gel cartridges by washing them with two 4-mL portions of 25% DCM in pentane; discard the washes.
18. Elute the nitrosamines with two 4-mL portions of 30% diethyl ether in DCM. Collect the eluate in 10-mL concentrator tubes.
19. Attach a micro-Snyder column to the concentrator tube and concentrate solvent to approximately 0.5 mL in a 70°C water bath. Dilute to a final volume of 1.0 mL with DCM.
20. Quantitate nitrosamines on GC-TEA for bacon or ham (*see* **Note 9**). Reported performance criteria for normally incurred nitrosamines in fried bacon ***(15)*** are NPYR, range, 0.7–20.2 ppb, mean 4.9 ppb, with a coefficient of variation (CV) of 4.1%; NDMA, range, none detected (ND)–2.4 ppb, mean 0.9 ppb, CV 12.6%; for nitrosamines in ham ***(14)***; NDBzA, range, ND-157.3 ppb, mean 63.2 ppb, CV 2.7%.
21. Total time to prepare duplicate samples for quantitation is about 1 h; GC-TEA analysis time is approximately 25 min.

4. Notes

1. Silica gel: The 70–230 mesh material was washed twice with DCM, filtered and dried for 4 h in a vacuum oven set at 60°C. It was sieved to a particle range of 70–150 mesh before use.
2. Hydromatrix: Sieved at 30–40 mesh to remove fine particles.
3. Caution: *N*-nitrosamines are potential carcinogens. Exercise care in handling these compounds. Store in amber bottles in a 4°C refrigerator when not in use, since the nitrosamines are photolabile.
4. Nitrosamines were synthesized from the corresponding amine and sodium nitrite as follows: cool an equimolar amine–hydrochloric acid solution with ice. Slowly, add a twofold excess of an aqueous solution of sodium nitrite to the amine–acid solution. After addition is complete, heat the reaction mixture at 60°C for 1 h. Extract the nitrosamine three times with diethyl ether. Dry the combined extracts over anhydrous sodium sulfate, then filter and concentrate under a stream of nitrogen. Distill the nitrosamine under vacuum ***(16)***.
5. Add the sample mixture to the extraction vessel in approximately four equal parts, compressing after each addition.
6. If there is more than a 1-cm space between the end of the compressed wool and the top of the extraction vessel, fill the space with additional polypropylene wool.
7. Add the silica gel to the cartridge followed by a polyethylene frit. Cut a 4-mm hole in another frit using a No. 1 cork borer and place the frit in the cartridge approximately 10 mm above the silica gel. This will prevent sample loss during decompression of the CO_2.
8. Set oven temperature initially to 43°C, then to 40°C after the vessels reach the desired temperature.

9. Analyze all samples in duplicate. Nitrosamines in the individual samples are corrected for recovery of the NDPA internal standard. Minimum levels of reliable measurement should have a signal-to-noise ratio of > 2.

References

1. Preussmann, R. and Stewart, B. W. (1984) *N*-Nitroso Compounds, in *Chemical Carcinogens* (Searle, C. E., ed.), ACS Monograph **182,** Vol. 2, 2nd ed., ACS, Washington, DC, pp. 643–828.
2. Tricker, A. R. and Preussmann, R. (1991) Carcinogenic *N*-nitrosamines in the dirt: occurrence, formation, mechanisms and carcinogenic potential. *Mutat. Res.* **259,** 277–289.
3. Fine, D. H., Rounbehler, D. P., and Oettinger, P. E. (1975) Rapid method for the determination of sub-part per billion amounts of *N*-nitroso compounds in foodstuffs. *Anal. Chim. Acta* **78,** 383–389.
4. Greenfield, E. I., Smith, W. J., and Malanoski, A. J. (1982) Mineral oil vacuum distillation method for nitrosamines in fried bacon with thermal energy analyzer: collaborative study. *J. Assoc. Offic. Anal. Chem.* **65,** 1319–1332.
5. Sen, N. P., Seaman, S. W., and Miles, W. F. (1979) Volatile nitrosamines in various cured meat products: effect of cooking and recent trends. *J. Agric. Food Chem.* **27,** 1354–1357.
6. Pensabene, J. W., Miller, A. J., Greenfield, E. I., and Fiddler, W. (1982) Rapid dry column method for the determination of nitrosopyrrolidine in fried bacon. *J. Assoc. Offic. Anal. Chem.* **65,** 151–156.
7. Pensabene, J. W., Fiddler, W., Gates, R. A., Fagan, J. C., and Wasserman, A. E. (1974) Effect of frying and other cooking conditions on nitrosopyrrolidine formation in bacon. *J. Food Sci.* **39,** 314–316.
8. Sen, N. P., Baddoo, P. A., and Seaman, S. W. (1987) Volatile nitrosamines in cured meats packaged in elastic rubber nettings. *J. Agric. Food Chem.* **35,** 346–350.
9. Sen, N. P., Seaman, S. W., Baddoo, P. A., and Weber, D. (1988) Further studies on the formation of nitrosamines in cured pork products packaged in elastic rubber nettings. *J. Food Sci.* **53,** 731–738.
10. Sen, N. P. (1991) Recent studies in Canada on the occurrence and formation of *N*-nitroso compounds in foods and food-contact materials. *IARC Sci. Publ.* **105,** 232–234.
11. Fiddler, W., Pensabene, J. W., Gates, R. A., Custer, C., Yoffe, A., and Phillipo, T. (1997) *N*-Nitrosodibenzylamine in boneless hams processed in elastic rubber nettings. *J. AOAC Int.* **80,** 353–358.
12. Fine, D. H., Rufeh, F., and Gunther, B. (1973) A group specific procedure for the analysis of both volatile and nonvolatile *N*-nitroso compounds in picogram amounts. *Anal. Lett.* **6,** 731–733.
13. Maxwell, R. J., Pensabene, J. W., and Fiddler, W. (1993) Multiresidue recovery at PPB levels of 10 nitrosamines from frankfurters by supercritical fluid extraction. *J. Chromatogr. Sci.* **31,** 212–215.

14. Pensabene, J. W., Fiddler, W., Maxwell, R. J., Lightfield, A. R., and Hampson, J. W. (1995) Supercritical fluid extraction of *N*-nitrosamines in hams processed in elastic rubber nettings. J. *AOAC Int.* **78,** 744–748.
15. Fiddler, W. and Pensabene, J. W. (1996) Supercritical fluid extraction of volatile *N*-nitrosamines in fried bacon and its drippings: method comparison. *J. AOAC Int.* **79,** 895–901.
16. Pensabene, J. W., Fiddler, W., Dooley, C. J., Doerr, R. C., and Wasserman, A. E. (1972) Spectral and gas chromatographic characteristics of some *N*-nitrosamines. *J. Agric. Food Chem.* **20,** 274–277.

4

Supercritical Fluid Extraction of Melengestrol Acetate from Bovine Fat Tissue

Robert J. Maxwell, Owen W. Parks, Roxanne J. Shadwell, Alan R. Lightfield, Carolyn Henry, and Brenda S. Fuerst

1. Introduction

Melengestrol acetate (MGA)—17α-hydroxy-6-methyl-16-methylene-pregna-4,6-diene-3,20-dione acetate (**Fig. 1**)—is a synthetic oral progestational steroidal hormone that is added to the feed of heifers to suppress estrus (heat), thereby leading to improved feed efficiency and rate of weight gain. In the United States, the Food and Drugs Administration (FDA) has set the tolerance level for residues of MGA in edible tissues at 25 ppb based on evidence that residues at or below this concentration do not elicit a hormonal response *(1)*, whereas in the European Union (EU) the residue limit for this steroid in animal products is 0 ppb *(2)*.

Several solvent extraction procedures are available for detecting MGA at or below the FDA tolerance level *(3–7)*. All of the reported methods use large amounts of organic solvents, many of which are halogenated. For instance, the method used by the Food Safety Inspection Service (FSIS) at the U.S. Department of Agriculture to detect MGA in bovine fat tissue requires 1.9 L of organic solvent per sample *(3)*. This is a matter of concern because the U.S. Environmental Protection Agency (EPA) has mandated that Federal laboratories and others reduce or eliminate the use of certain organic solvents *(8)*. Hence solvent-sparing technologies must be investigated to determine their suitability for regulatory laboratories.

Mention of brand or firm names does not constitute an endorsement by the US Department of Agriculture over others of a similar nature not mentioned.

From: *Methods in Biotechnology, Vol. 13: Supercritical Fluid Methods and Protocols*
Edited by: J. R. Williams and A. A. Clifford © Humana Press Inc., Totowa, NJ

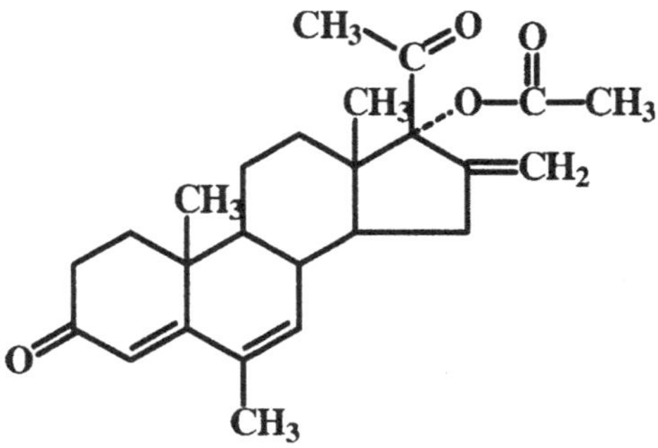

Fig. 1. Chemical structure of melengestrol acetate.

The fundamental principles of supercritical fluid extraction (SFE) have been covered in Chapter 1. Extraction with supercritical fluids has been used by others to isolate steroids such as androsterone from boar fat ***(9)***. In that study, the steroid was collected *off-line* [after carbon dioxide (CO_2) decompression] together with coextracted fat. This method of analyte collection requires several post-SFE clean-up operations to separate the androsterone from coextracted fat prior to chromatographic analysis. Maxwell et al. ***(10)*** developed an alternative technique to *off-line* analyte collection where three steroids, nortestosterone, testosterone, and methyltestosterone, were trapped on an alumina sorbent bed under supercritical fluid conditions (*in-line* trapping). This technique is illustrated in **Fig. 2**, which shows an SFE vessel prepared for in-line analyte collection. Analytes such as steroids are retained on the *in-line* sorbent bed while fat and other fat-soluble coextractables are deposited in an *off-line* vial after CO_2 decompression thereby eliminating the need for multiple post-SFE clean-up operations.

This chapter describes a method for the SFE of MGA from bovine fat tissue using *in-line* trapping. Because of the solvent intensive nature of the current methods for MGA, the *in-line* analyte collection technique was employed for the recovery by SFE of MGA from bovine fat tissue (**Fig. 2**). Unlike the official FSIS method, the SFE MGA method requires only a single post-SFE solid-phase extraction clean-up step prior to chromatographic analysis and consumes only 12 mL of methanol. Recoveries of MGA from fortified tissues were $98.4 \pm 4.5\%$ at the 25 ppb level. **Table 1** shows calculated concentration values of incurred residues of MGA from bovine fat tissues that compared favorably to those obtained by the FSIS procedure ***(11)***. Chromatograms [derived from high performance liquid chromatography (HPLC) with ultraviolet (UV) detection] of control and incurred fat samples indicate that MGA can be quantified easily by the SFE method at or below the 25 ppb level without interference

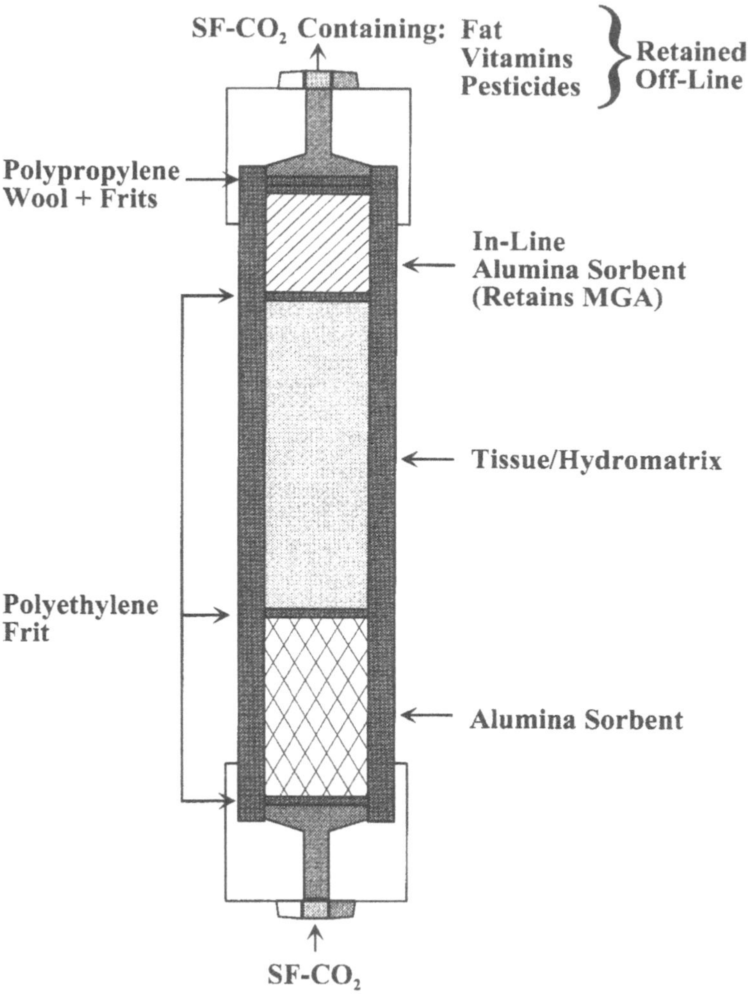

Fig. 2. Schematic drawing of high pressure extraction vessel showing layering of *in-line* trap, sample mixture and presample trap.

from UV-absorbing background material (**Fig. 3**). Confirmation of MGA in the incurred samples was determined by GC-MS of the HFB enol ester derivative. The total selected ion current chromatogram and selected ion current profiles of an MGA-HFB standard are shown in **Fig. 4A**, while **Fig. 4B** shows a total selected ion current chromatogram of a control extract and the selected ion current chromatograms from an incurred fat extract. Note that the total

Table 1
Concentration of Incurred Residues of Melengestrol Acetate in Bovine Fat Tissue as Determined by Organic Solvent* and SFE Procedures

		Concentration (ppb ± SD)	
Fat sample	Animal number	Solvent (n = 3)	SFE (n = 5)
Visceral	6004	20 ± 4.7	24.8 ± 1.1
Perirenal	6028	57 ± 6.0	53.9 ± 1.1
Visceral	6036	85 ± 14.5	89.4 ± 4.2
Perirenal	6036	108 ± 6.6	97.7 ± 4.6

*Food Safety Inspection Service (*see* ref. ***3***). Reproduced from the *Journal of Chromatographic Science* by permission of Preston Publications, A Division of Preston Industries, Inc. ppb, parts per billion; SD, standard deviation; n, the number of determinations.

selected ion current profile of the control extract (**Fig. 4B**) was void of peaks in the retention windows for the molecular ion of MGA-HFB and its characteristic fragment ions.

2. Materials

1. Two high-pressure vessels (10,000 psi, 24-mL capacity, Keystone Scientific, Bellefonte, PA) are extracted in parallel with the use of the Spe-ed SFE Model 680 bar extraction system (Applied Separations, Allentown, PA). The SFE apparatus is equipped with a thermocouple to monitor extraction vessel temperature. The air-driven Haskel pump contained in the system is equipped with a chiller cooled by a refrigerated circulating bath set at –15°C. The use of this device obviates the need for helium-pressurized CO_2, which is required for standard operation with a noncooled pump-head.
2. The extracted fat is collected *off-line* in 9-mL vials fitted with septa. The vials are vented to a Floline SFE-51 flow meter/gas totalizer (precalibrated for CO_2 gas and purchased from Scott Specialty Gases, Plumbsteadville, PA).
3. Hydromatrix or Celite 566 (part no. 0019–8003 Varian Sample Preparation Products, Harbor City, CA).
4. Alumina (Al_2O_3)—activated, neutral, Brockmann I (catalog no. 19,997.4 Aldrich Chemical Co., St. Louis, MO), used as received.
5. Solid-phase extraction (SPE) columns (6 mL) containing 1.0 g 18% C_{18} packing (Applied Separations).
6. Methanol (MeOH), acetone, ethyl acetate (EtOAc), isooctane, and acetonitrile (CH_3CN) are high-purity solvents.
7. Supercritical fluid chromatography-grade CO_2 with a diptube and no helium headspace (Scott Specialty Gases).
8. Polypropylene wool from Aldrich Chemical Co. (*see* **Note 1**).
9. Tamping rod (~12 mm diam.) and polyethylene frits of 35 μm pore size (*see* **Note 2**).

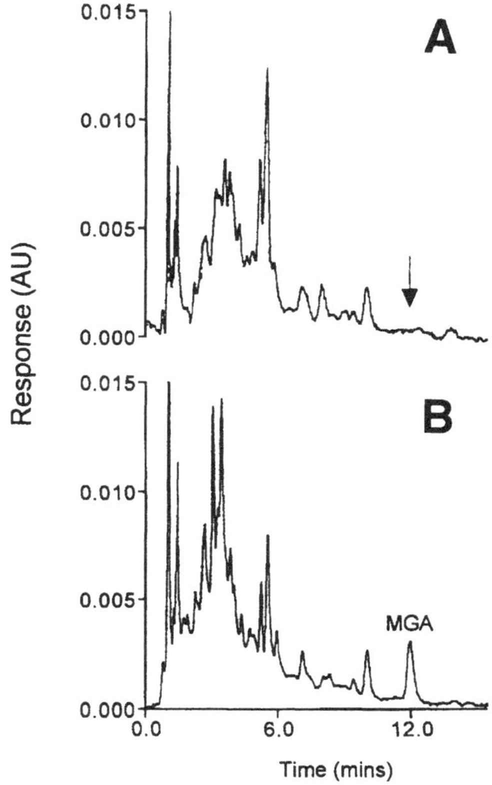

Fig. 3. HPLC chromatograms of supercritical CO_2 extracts of **(A)** control sample of perirenal fat tissue and **(B)** visceral fat tissue (animal number 6004) containing incurred residues of melengestrol acetate or MGA (reported concentration, 20 ppb; determined concentration, 24.8 ppb). [Reproduced from the *Journal of Chromatographic Science* by permission of Preston Publications, A Division of Preston Industries, Inc.]

10. Heptafluorobutyric acid anhydride or HFBA (cat. no. 63164 Pierce Co, Rockford, IL).
11. Microreaction vessel (cat. no. 3-3291 Supelco, Bellefonte, PA).
12. Melengestrol acetate (MGA) is a control reference standard of the Upjohn Company, Kalamazoo, MI (*see* **Note 3**).
13. Samples of bovine perirenal and visceral fat tissues containing varying levels of MGA are obtained from the USDA, FSIS Midwestern Laboratory (*see* **Note 4**).
14. HPLC: Isco (Lincoln, NE) LC-5000 syringe pump equipped with a Rheodyne (Berkeley, CA) Model 7125 injector connected to a Supelcosil LC-18 column (15 cm × 4.6 mm ID, 5-µm particle size by Supelco). MGA is detected at 291 nm with an Applied Biosystems (Foster City, CA) Model 1000S UV diode array detector. The mobile phase is $CH_3CN:H_2O$ (55:45, v/v) at a flow rate of 1.0 mL/min.

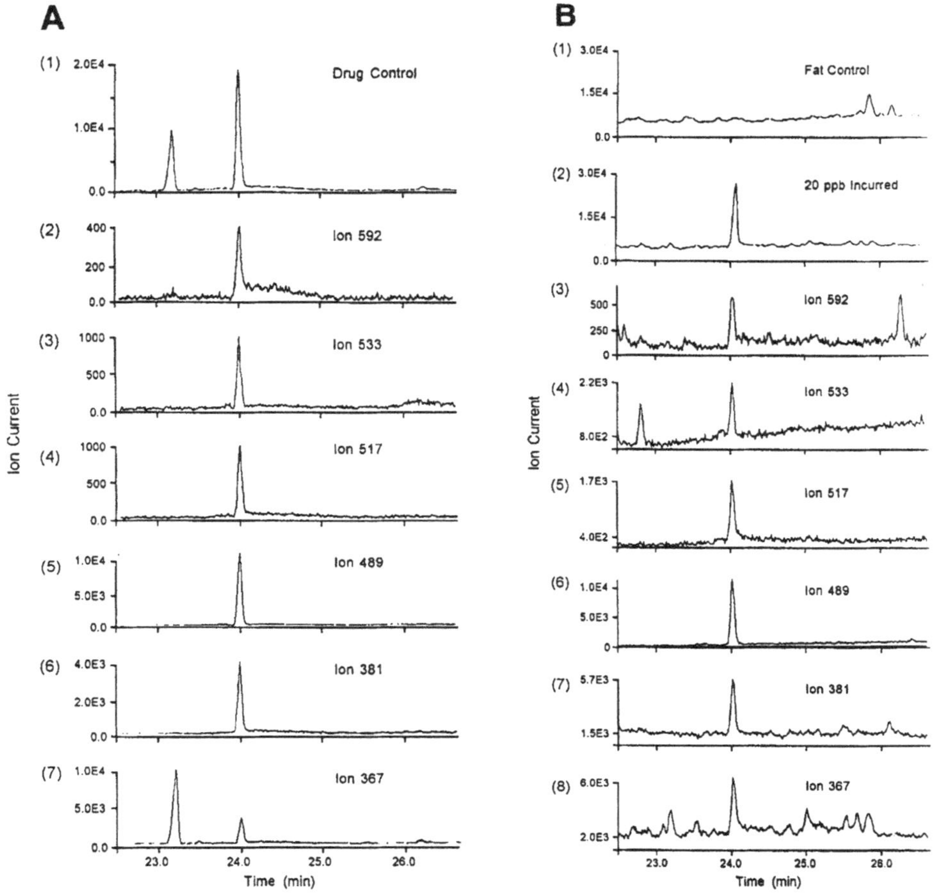

Fig. 4. **(A)** GC-MS profiles of total (No. 1) and individual (No. 2-7) selected ion currents of a melengestrol acetate–heptafluorobutyric acid (MGA-HFB) standard (equivalent to 25 ppb) (t_r, 24 min). **(B)** Total selected ion current GC-MS profiles of control fat (No. 1) and MGA incurred fat tissue (No. 2) extracts and the individual selected ion current profiles (No. 3–8) of the incurred tissue extract (visceral fat; animal no. 6004). [Reproduced from the *Journal of Chromatographic Science* by permission of Preston Publications, A Division of Preston Industries, Inc.]

Chromatograms are recorded on a Hewlett-Packard (HP, Avondale, PA) Model 3396A integrator. Quantitation of MGA is accomplished by comparison of peak heights or areas (or both) with external standards.

15. Gas chromatography–mass spectrometry (GC-MS) analysis is performed according to the procedure of Chichila and coworkers *(7)* using HP Model 5890 GC equipped with an HP Model 7673 GC auto injector and an HP GC autosampler

interfaced to an HP Model 5970 mass selective detector. The capillary column is a crosslinked methylsilicone gum (HP ultra-1, 12 m × 0.22 mm ID × 0.3 mm film thickness, HP no. 109091A-101). The injector temperature is maintained at 260°C, and the interface temperature is 300°C. The oven temperature is set at 40°C, programmed at 30°C/min to 150°C, and then at 6°C/min to 300°C. The final temperature is held for 10 min. The presence of the 3-heptafluorobutyrylenol ether of MGA (MGA-HFB) is confirmed by selected ion current monitoring for the molecular ion (m/z 592) and five characteristic fragments (m/z 533, 517, 489, 381 and 367) and their total absence in control fat tissue extracts (**Fig. 4**).

3. Method

1. Place 1.0 g of a rectangular slice of negative control perirenal fat tissue on a watch glass and fortify with 3 µL of the MGA fortification solution in a standard 10 µL syringe by depositing the solution on the surface of the tissue (*see* **Note 5**).
2. Add the fortified tissue to 4.0 g of Hydromatrix contained in a 50-mL beaker, then add dropwise 0.75 mL of distilled H_2O.
3. Grind the tissue thoroughly into the "wetted" Hydromatrix with a metal spatula.
4. Cap and seal one end of an SFE high pressure vessel and label that end top.
5. Pack the extraction vessel tightly (*see* **Note 6**) in the following sequence relative to the top of the vessel: a plug of polypropylene wool, two polyethylene frits, 2 g of neutral alumina (analyte trap), a polyethylene frit, fortified or incurred tissue-Hydromatrix mixture (dry, free-flowing sample mixture), a polyethylene frit, 3 g of alumina (presample trap - *see* **Note 7**) and a polyethylene frit (**Fig. 2**). Cap bottom end of vessel.
6. The SFE inlet, outlet and vent valves should be closed and the micrometering valves set to a minimum flow rate. Install the packed extraction vessels in the SFE oven with the end labeled top connected to the upper slip-free fittings and attach the built-in thermocouple to one extraction vessel (*see* **Note 8** and Chapter 3, **Fig. 1**).
7. Attach a 9-mL vial to each micrometering valve *off-line* interface for fat collection. The flow rate and total CO_2 (expanded gas) are monitored with a flowmeter/gas totalizer alternately connected to each *off-line* vial.
8. Preheat the micrometering valves to 120°C.
9. Set the oven temperature to 50°C and begin heating.
10. When the vessel set point temperature is reached, open the inlet valves and increase the pump pressure to 10,000 psi or 680 bar (*see* **Note 9**).
11. Equilibrate the system with a 5 min static holding period.
12. After 5 minutes, open the outlet valves. Then slowly adjust each micrometering valve flow rate to 2 L/min (expanded gas) for each vessel.
13. After 40 L are recorded by the totalizer, close the inlets valves and depressurize the extraction vessels under controlled flow conditions using the micrometering valves (*see* **Note 10**).
14. Attach an empty 6-mL SPE column fitted with a polyethylene frit to a stand. Directly below this column attach a 6 mL SPE column containing 1.0 g of 18% C_{18} packing. Set aside until **step 17**.

15. After extraction vessel decompression, remove the vessel(s) from the SFE oven and uncap the end labeled top (*see* **Note 11**). Remove and discard the polypropylene wool and frits.
16. Carefully pour the vessel's alumina sorbent layer into the empty 6 mL SPE column (*see* **step 14**). Compact the sorbent by tapping the sides and top of the SPE column with a spatula, then layer the top of the sorbent bed with 0.25 cm of clean sand.
17. Elute the SPE column with MeOH/H_2O (65:35 v/v). Allow the first 2 mL of eluate to pass into the C_{18} SPE column below.
18. Wash the C_{18} SPE column containing the MeOH/H_2O eluate with two 1-mL portions of MeOH/H_2O (65:35 v/v) and two 2-mL portions of deionized water.
19. Dry the C_{18} SPE column by vacuum and elute with MeOH. Collect 2 mL of the eluant from this SPE column in a 5 mL screw-capped vial.
20. Evaporate the MeOH in the vial to dryness under a nitrogen stream.
21. For HPLC analysis, *see* **Subheading 2, step 14** and add 250 µL of the HPLC mobile phase to the contents of the vial and vortex for 30 s. Draw up 100 µL of the resultant solution in a syringe and inject into the HPLC.
22. For GC-MS analysis, *see* **Subheading 2, step 15** and first prepare the HFBA derivative of MGA *(3,11)* by collecting 2-mL of MeOH eluant from the C_{18} SPE column (*see* **step 19**) in a 2 mL Teflon-lined screw-capped vial and evaporate to dryness under a nitrogen stream. Add 80 µL of acetone and 20 µL of HFBA to the residue. Vortex the mixture for 1 minute and then heat at 60°C for 1 h. Transfer contents of vial to a 0.3 mL micro Supelco reaction vessel. Rinse the transfer vial with 100 µL of acetone and add that to the contents of the reaction vessel. Evaporate the contents of the vessel to dryness at room temperature under a nitrogen stream. Take up the residue in 10 µL of EtOAc-isooctane (5:95 v/v) and seal the vessel with a cap fitted with a septum. Vortex the vessel and centrifuge. Inject 3 µL of the solution into the GC-MS.
23. Quantitate MGA by HPLC or GC-MS. Performance criteria for normally incurred MGA in bovine fat tissue are shown in **Table 1**.
24. Total time to prepare the sample for quantitation is approximately 1 h.

4. Notes

1. Preclean polypropylene wool by compressing an amount to fill a 24 mL high pressure vessel and extracting the wool for 20 min at 10,000 psi (680 bar), 50°C and a CO_2 flow rate of 3 L/min (expanded gas).
2. Inexpensive polyethylene frits for SFE extraction vessels can be made in the laboratory by punching disks from 35 µm porous polyethylene sheets (Bel-Art Products, Pequannock, NJ) using a number 8 stainless steel cork hole borer.
3. Fortification solutions containing 34, 17, and 8.5 ng/mL of MGA in MeOH were prepared and used to fortify tissue samples.
4. These samples were extruded through a meat grinder and were analyzed for MGA by FSIS, USDA using their official solvent extraction procedure *(3)*.
5. Hold the fortified tissue at room temperature for 10 min before beginning **step 2** in order to allow permeation of MGA into the tissue and for evaporation of the MeOH.

6. Compress the material tightly in the vessel with the tamping rod after adding each successive layer. Refrigerate the packed vessel to prevent analyte loss if it is not to be immediately extracted by SFE.
7. The purpose of the presample trap is to prevent any contaminants from the SFE pump or the CO_2 cylinder from reaching the *in-line* analyte trap.
8. The vessel temperature is monitored separately from the oven temperature in order to ensure reproducible analyte recovery.
9. Monitor vessel temperature on thermocouple display not oven temperature readout to ensure that vessel temperature does not exceed the set point during vessel pressurization.
10. Do not use the vent valves to depressurize the system.
11. It is neither necessary to clean the transfer lines from the SFE vessel to the micrometering valves after each use, nor is it required to replace the *off-line* fat collection vials on a daily basis. However, in the event that the transfer lines are to be cleaned, attach Luer adapters to the upper slip-free connectors in the oven and attach a 1-mL syringe filled with 0.3 mL of hexane to each adapter. Flush fat residues in transfer lines into the *off-line* collection vials.

References

1. Anonymous (1994) Melengestrol acetate clearances broadened. *Food Chem. News*, August 15, p. 34.
2. Heitzman, R. J. (1992) *Agriculture Veterinary Drug Residues in Food-Producing Animals and Their Products: Reference Materials and Methods.* Commission of the European Communities Monograph, Brussels, Luxembourg, M. 1. 1.
3. Food Safety and Inspection Service (1991) *Analytical Chemistry Laboratory Guidebook: Residue Chemistry 5.040.* United States Department of Agriculture, Washington, D.C.
4. Food and Drug Administration, Department of Health and Human Services (1993) *Code of Federal Regulations, 21 C.F.R. 556.380.* U.S. Government Printing Office, Washington DC.
5. Association of Official Chemists (1990) *Official Methods of Analysis,* 14th ed. Association of Official Analytical Chemists, Washington, D.C., pp. 629–631.
6. Ryan, J. J. and Dupont, J. A. (1975) Measurement and presence of melengestrol acetate (MGA) in beef tissues at low levels. *J. Agric. Food Chem.* **23,** 917–920.
7. Chichila, T. M. P., Edlund, P. O., Menion, J. D., Wilson, R., and Epstein, R. L. (1989) Determination of melengestrol acetate in bovine tissues by automated coupled-column normal phase high performance liquid chromatography. *J. Chromatogr.* **488,** 389–406.
8. U.S. E.P.A. (1991) Fed. Reg., Vol. 56: *U.S. E.P.A. Pollution Prevention Strategy.* U.S. E.P.A., Washington, D.C., pp. 7849–7864.
9. Mågård, M. A., Berg, H. E. B., Tagesson, U., Järemo, M. L. G., Karlsson, L. L. H., Mathiasson, L. J. E., Bonneau, M., and Hansenn-Moller, J. (1995) Determination of androsterone in pig fat using supercritical fluid extraction and gas chromatography-mass spectrometry. *J. Agric. Food Chem.* **43,** 114–120.

10. Maxwell, R. J., Lightfield, A. R., and Stolker, A. A. M. (1995) An SPE column-Teflon sleeve assembly for in-line retention during supercritical fluid extraction of analytes from biological matrices. *J. High Resol. Chromatogr.* **18,** 231–234.
11. Parks, O. W., Shadwell, R. J., Lightfield, A. R., and Maxwell, R. J. (1996) Determination of melengestrol acetate in supercritical fluid-solid phase extracts of bovine fat tissue by HPLC-UV and GC-MS. *J. Chromatogr. Sci.* **34,** 353–357.
12. Stolker, A. A. M., van Ginkel, L. A., Stephany, R. W., Maxwell, R. J., Parks, O. W., and Lightfield, A. R. (1999) Supercritical fluid extraction of nortestosterone, testosterone and methyltestosterone at low ppb levels from fortified bovine urine. *J. Chromatogr. B* **726,** 121–131.

5

Supercritical Fluid Extraction of Polychlorinated Biphenyls from Fish Tissue

Michael O. Gaylor and Robert C. Hale

1. Introduction

Polychlorinated biphenyls (PCBs) are of great concern to the scientific and regulatory communities due to their tendency to accumulate to toxic levels in the edible tissues of fish and other organisms *(1–4)*. PCBs are nonpolar compounds that can partition into the lipid reservoirs of edible tissues causing damage to ecosystems and human health *(5,6)*. Despite significant progress in environmental reform, extraction methodologies required to isolate PCBs continue to rely heavily on environmentally deleterious liquid organic solvent extraction methods such as Soxhlet extraction, sonication, and column elution *(7–9)*. These techniques are laborious, tedious, analyte-nonselective, and require copious volumes of organic solvents. Common solvents are typically toxic or flammable and ultimately must be disposed of as hazardous waste. Traditional solvent extracts obtained require multiple postextraction purification steps, such as gel permeation chromatography (GPC), florisil, and silica column clean-up *(10)*. These steps contribute further to the hazardous waste disposal problem facing environmental laboratories. The entire process is paradoxical in that it contradicts the intended goal of these procedures, that of improving environmental quality.

By contrast, supercritical fluid extraction or SFE (Chapter 1) has emerged in recent years as a more environmentally benign analytical technique that promises to significantly improve the extraction of trace organic pollutants, such as PCBs, from environmental samples *(11,12)*. The practical advantages of SFE for PCB determinations in environmental samples include minimal sample manipulation, rapid extractions (30–60 min), improved analyte selectivity and

From: *Methods in Biotechnology, Vol. 13: Supercritical Fluid Methods and Protocols*
Edited by: J. R. Williams and A. A. Clifford © Humana Press Inc., Totowa, NJ

recovery, no postextraction clean-up, enhanced automation potential, and a drastic reduction in liquid solvent usage.

The vast majority of environmental research has focused on SFE of abiotic matrices such as soils, sediments, sludges, and fly ash *(13–15)*. Comparably little research has been conducted in applying SFE to trace-level organic pollutant determinations in aquatic biota samples *(11,16,17)*. The extremely high water content of aquatic organisms (80–90%) and appreciable tissue lipid solubility in supercritical carbon dioxide have been major deterrents to progress on this front. Complete removal of water from the sample is critical for SFE because of the potential to freeze and plug the restrictor and cryogenic trap during extraction. Further, because of the negligible miscibility of supercritical phase CO_2 and water (< 0.1% w/w) under a given set of temperature and pressure conditions, sample water can interfere with analyte/solvent interactions, preventing analyte dissolution in the extraction solvent *(18,19)*. Water can also alter the critical parameters of the extraction solvent, leading to diminished extraction efficiency *(20)*. Numerous preextraction chemical desiccation approaches have been used for abiotic matrices, including diatomaceous earth (i.e., Hydromatrix), sodium sulfate, calcium chloride, magnesium sulfate, alumina, and florisil *(21)*. However, these materials can occupy significant internal vessel volume and may solidify upon reaction with water, leading to undesirable effects such as reduced sample size and concomitant increases in analyte-detection limits, loss of water from drying agents at elevated temperatures, and plugged extraction vessels. Recent studies have demonstrated the feasibility of retaining coextracted lipids during SFE of biological samples by adding alumina directly to the extraction vessel *(11,16,17)*. Lipid-free sample extracts eliminate the need for GPC and polarity-based purification, promote quality chromatographic separations and prolong the operating performance of gas chromatograph injector ports and analytical columns. Obtaining extracts that are as free as possible of coextracted lipids should, therefore, be a high priority when developing SFE methods for any biological matrix.

To address the lack of data in this important area of environmental research, a simple protocol for the determination of PCBs in freeze-dried edible fish tissue using off-line SFE is presented. The method is rapid, requiring only 40 min per dry sample and is amenable to automation. The addition of activated neutral alumina directly to the top of the sample during SFE retains greater than 99% of coextractable lipids, eliminating completely the need for postextraction clean-up. After SFE, PCBs are desorbed into 2-mL gas chromatograph autosampler vials with 1.8 mL of isooctane, thus reducing total solvent consumption by as much as two orders of magnitude per sample. The extracts can be assayed directly using gas chromatography with electrolytic conductivity detection (GC-ELCD) in the halogen-selective mode *(22)*. The method saves considerable time (hours vs days) and solvent (milliliters vs liters) compared to conventional liquid solvent-based techniques and is capable of selectively extracting PCBs from fatty fish tissue samples.

2. Materials

1. Supercritical fluid extractor (AP44™, Isco-Suprex Inc., Lincoln, NE); solid-phase, cryogenic trapping unit (Accutrap™, Isco-Suprex Inc.); 10 mL stainless steel extraction vessels; porous PEEK vessel frits; frit crimping wrench (Isco-Suprex Inc.).
2. Freeze-dryer (Dura-Dry model, FTS Systems, Inc., Stony Ridge, NY).
3. Analytical balances (Mettler, Hightstown, NJ; Ohaus, Florham, NJ, *see* **Note 1**).
4. (a) Surrogate PCB standard(s) diluted in hexane, containing: 1) IUPAC congeners 30, 65, and 204 or 2) PCB congeners ranging in degree of chlorination from mono- to decachlorobiphenyl (b) An internal quantitation standard (i.e., pentachlorobenzene, PCB 204 or PCB 207; Ultra Scientific, Kingstown, RI, *see* **Note 2**).
5. Organic solvents (hexane, isooctane, benzene, *n*-propyl alcohol, acetone, methylene chloride, methanol) certified for pesticide residue trace analyses.
6. Ultrahigh purity helium and hydrogen (minimum purity 99.999%) for GC-ELCD analysis; prepurified nitrogen (minimum purity 99.995%) for purging residual solvent and analytes from the cryogenic trap after desorption, actuation of pneumatic valves on the AP44™ and Accutrap™ units, solvent evaporation before GC analysis and GC autosampler operation; scientific-grade nitrogen (minimum purity 99.999%) for freeze- drying; industrial-grade CO_2 for cooling the cryogenic trap during SFE; ultrahigh purity SFE/SFC-grade CO_2 with at least 10.2 MPa (102 atm) helium head for sample extraction. SFE/SFC-grade CO_2 should conform to the following purity specifications: < 2 ppm hydrogen, < 20 ppm nitrogen, < 2 ppm oxygen, < 2 ppm carbon monoxide, < 0.5 ppm water and total ECD and FID response < 100 ppt and 2 ppb, respectively (Air Products, Hampton, VA; Scott Specialty Gases, Plumsteadville, PA; MG Industries, Malvern, PA).
7. C_{18}-modified silica, 30 µm (Aldrich Chemical, Saint Louis, MO); 80/100 mesh (60 Å pore size) Unibeads 2S modified silica and 100/120 mesh silanized glass beads (Alltech, Deerfield Park, IL); 150 mesh activated neutral alumina (50 Å pore size, Brockmann 1 activity, 155 m^2/g surface area) for use in the solid-phase trap.
8. GC/HPLC vials (2/12 mL) equipped with plastic screw caps and Teflon-lined septa; TurboVap LV solvent evaporator (Zymark Inc., Hopkinton, MA) for solvent extract collection and sample volume reduction (*see* **Note 3**).
9. Stainless steel spatula, freeze-drying and sample storage pans; glass rod for sample and trap compaction prior to SFE, fillet knife and glass fillet board.
10. Safety equipment: latex gloves for washing and solvent rinsing all sample contact surfaces (i.e., extraction vessels, sample jars, fillet board, etc.) and filleting fish samples.
11. Model 3400 gas chromatograph (Varian, Walnut Creek, CA) equipped with a Model 4420 electrolytic conductivity detector (OI Corporation, College Station,

TX), 60 m DB-5 fused silica capillary column (J & W Scientific, Folsom, CA; 0.32 mm inner diameter and 0.25 µm film thickness); Model 8100 GC autosampler (Varian); GC gas purification filters (MG Scientific, Malvern, PA); Model 3350 Laboratory Automation System (LAS) computer data acquisition system and Model 35900 A/D signal converter (Hewlett Packard, Palo Alto, CA); Model ELQ 400-2 negative chemical ionization mass spectrometer (Extranuclear Corp., Pittsburgh, PA, *see* **Note 4**).

3. Method

1. Glassware cleaning: clean all glassware and other surfaces that will contact the sample with laboratory-grade detergent (Alconox) followed by soaking in a 10% solution of Contrad 70 (Curtin Matheson Scientific, Atlanta, GA) in deionized water for a minimum of 4–6 h *(23)*. Soaking overnight is preferred. Remove glassware from Contrad 70 solution, rinse with deionized water and allow to air dry. Bake volumetric items overnight in an oven at 100°C. Bake nonvolumetric items for 4–6 h at 400°C. Before preparing samples, rinse all sample contact surfaces with a suite of organic solvents ranging in polarity from moderately polar to nonpolar. A typical sequence is methanol, acetone, methylene chloride, and hexane.
2. Edible fish tissue sample handling and preparation: immediately after collection, wrap the fish in solvent-rinsed aluminum foil, pack on ice, and transport to the laboratory. Remove edible fillet tissue and place in a clean, preweighed stainless steel freeze-dryer pan (*see* **Note 5**). Reweigh the pan and wet sample to determine percent moisture after freeze-drying. Cover the samples with aluminum foil and freeze overnight in preparation for drying.
3. Freeze-drying samples: rinse the freeze-dryer thoroughly with a methanol-soaked, lint-free disposable towel and allow it to completely dry before introducing samples. Remove sample pans from the freezer and place them immediately into the freeze-dryer. Peel back one corner of the foil to allow complete sublimation of sample water during freeze-drying. Freeze-dry at 0°C under a 600 mtorr vacuum. During freeze-drying, a positive pressure of nitrogen (0.5 MPa, ~5 atm) is provided to the freeze-dryer chamber to prevent pump oil from back-streaming and contaminating the samples. Samples typically require 24–48 h to dry thoroughly (*see* **Note 6**).
4. Sample homogenization: after drying, store foil-covered samples in a desiccator. Place each individual sample separately into a blender and homogenize at high speed until a powderlike consistency is achieved (*see* **Note 7**).
5. Activation of neutral alumina before SFE: Activate neutral alumina by pouring a 2- to 3-cm layer of alumina into a clean stainless steel freeze-drying pan or Pyrex dish. Heat overnight in a clean oven at 130°C.
6. Preparation of surrogate and internal standards: prepare surrogate standard(s) by dissolving known amounts of PCB congeners 30, 65, and 204 in hexane in a clean volumetric flask. An internal standard should be chosen and prepared similarly for use in quantitating PCBs in the sample (*see* **Note 8**).

7. Pre-SFE sample preparation: remove tissue samples from the freezer and allow them to warm to room temperature in the sample jars. Solvent rinse stainless steel extraction vessels (10 mL; *see* **Subheading 3., step 1**). Allow vessels to completely dry for several minutes under a fume hood. After drying, cap one end of the vessel (entrance/bottom) with a PEEK frit, seal with a crimping wrench and label the vessel with an indelible marker (*see* **Note 9**). Tare the vessel on an analytical balance. Place a small, clean glass or stainless steel funnel in the open end (exit/top) of the extraction vessel. Introduce the sample into the vessel using a clean spatula by gently scraping tissue from the jar and guiding it into the funnel in small amounts. Compact the tissue gently using a clean glass rod at regular intervals so that a homogeneous "plug" is formed. Remove any spilled sample material from the vessel rim and weigh at periodic intervals until the desired sample weight is achieved (usually 1 g). The end result should be a gently compacted, homogeneous "plug" of tissue in the bottom of the vessel. Again, remove excess sample material from the top rim spilled during vessel filling before recording the final sample weight (*see* **Note 9**). Using a graduated pipette, add the desired amount of surrogate standard directly to the top of the sample to assess the efficacy of the technique and account for procedurally related analyte losses. Allow carrier solvent to evaporate before continuing (*see* **Note 10**).
8. Addition of neutral alumina: remove alumina from the oven and transfer to a clean 250- to 500-mL beaker, cover with clean aluminum foil and allow to cool to room temperature in a desiccator (*see* **Note 11**). Once cooled, slowly pour the alumina directly into the exit end of the extraction vessel, on top of the sample, until the vessel is filled completely (*see* **Fig. 1**). Gently tamp the vessel periodically during alumina addition to compact the sorbent and eliminate voids. The final sorbent level should be ca. 0.2 cm below the vessel opening. Completely remove excess sorbent from the rim of the vessel opening (*see* **Note 9**). Cap the vessel with a PEEK frit and seal with the crimping wrench. Load the vessels into the SFE sample carousel.
9. Preparation of the cryogenic trap: disassemble the trap by removing both end caps and freeing the stainless steel center piece (*see* **Note 12**). If the trap has been used previously for a different suite of analytes, and contains sorbent incompatible with PCB trapping, blow out this material into an appropriate disposal receptacle using compressed air. Solvent rinse the trap to remove residual material from the inner surface. Cap the bottom end (exit end) and insert a small plug of glass wool into the top end (entrance end), compressing it to the bottom with a clean spatula or glass rod to retain the trapping sorbent during analyte collection. Fill the trap 3/4 full with a 1:1 (w/w) mixture of C_{18}-modified silica/Unibeads. Cap the top end and reattach the assembly to the Accutrap^{TM} module. Rinse the trap before SFE with 5- to 10-mL of isooctane at 1 mL/min to remove residual impurities and packing fines (*see* **Note 12**).
10. Sample extraction: enter the desired extraction parameters into the SFE unit using the key pad on the front of the instrument. The optimum parameters for extraction of PCBs from fish tissue with this configuration are: 10 min initial static

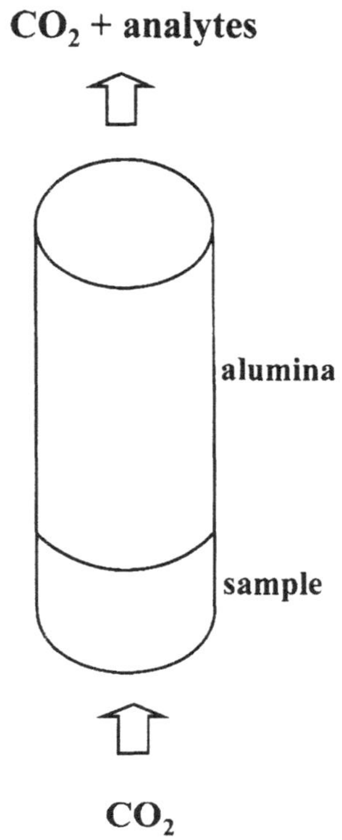

Fig. 1. Diagram of an SFE vessel showing the orientation of sample, alumina and direction of CO_2 flow and dissolved analytes during extraction.

extraction at 35 MPa (350 atm) and 150°C, followed by a 30 min dynamic extraction step at 35 MPa and 150°C with a compressed CO_2 flow rate of 3 mL/min (measured at the pump). The analytes are collected on the trap at 0°C. The restrictor is maintained at 100°C to eliminate freezing, caused by Joule-Thompson cooling during CO_2 expansion. After dynamic extraction, the trap is heated ballistically to 90°C and the analytes desorbed into a 2 mL GC autosampler vial with 1.8 mL of isooctane at a flow rate of 1 mL/min. After desorption, the remaining isooctane and analytes are purged from the trap with nitrogen. This prevents analyte carry-over between collection vials and promotes quantitative analyte recovery (*see* **Note 13**).

11. Preparation of SFE extract for GC-ELCD analysis: remove samples from the SFE fraction collector and reduce to the desired volume (i.e., 0.2–0.3 mL) under a gentle stream of nitrogen directly in the vial. Amend the extract with internal standard(s) before chromatographic analysis for use in quantitation of sample PCBs (*see* **Note 14**).
12. Analysis of SFE extract using GC-ELCD: 1–2 µL of extract are injected in the splitless mode (injector split vent opens after 2 min). Helium is used as the carrier

gas at a flow rate of 1 mL/min. The injector is maintained at 300°C and the ELCD at 900°C. The column temperature is held at 90°C for 2 min, programmed to a final temperature of 320°C at 4°C/min and held at 320°C for 10 min.

13. Compound identification and quantitation: PCBs are identified using a halogen retention index or HRI *(23)*. Sample PCBs are quantified using relative response factors of known individual PCB congeners. Response factors are determined by comparing the response of the internal standard to those of PCB congener standards using GC-ELCD. After quantitation, PCB concentrations in the sample are typically normalized to the recovery of surrogate compounds. Compound identification may be confirmed using GC with negative chemical ionization mass spectrometry or GC/NCI-MS (*see* **Subheading 3., step 12** for GC configuration). Methane is used as the moderator gas and the ion source temperature is maintained at 100°C under a 700-mtorr vacuum.
14. Quality assurance or quality control: continuously monitor quality assurance and control by extracting spiked blank matrices interspaced between real samples to assess analyte carryover, laboratory contamination and recovery of surrogate compounds in all samples. Spiked blanks can also be used to establish analyte solubility under a given set of extraction conditions and ensure that quantitative recoveries of surrogate compounds are obtained in the absence of matrix effects. Extract sample replicates and standard reference materials (SRMs) periodically to certify accuracy and precision of the protocol. Inject PCB standards containing congeners representing all degrees of chlorine substitution (i.e., mono-deca) at known concentrations daily to verify GC-ELCD and GC/NCI-MS system response.

4. Notes

1. Balances are required that are capable of weighing neat standards (mg), sample material (g), and stainless steel extraction vessels (>100 g). Two balances were used for this work, one high weight range for sample and extraction vessel weighing (Mettler) and the other for standard(s) preparation (Ohaus).
2. PCBs 30, 65, and 204 have been used extensively as surrogate standards during development of this method. They are consistently baseline-resolved in the presence of native PCBs during GC. Other congeners are potentially suitable provided they are also absent from commercial Aroclor mixtures, thus not occurring in environmental samples *(24)*. Recently, considerable SFE optimization work has been completed using a PCB by-product standard containing PCB congeners 1, 3, 7, 30, 50, 97, 143, 183, 202, 207, and 209. These compounds have proven valuable for assessing extraction efficiency as a function of both molecular weight and degree of chlorination from spiked blanks and real-world samples containing minimal incurred PCBs (i.e., 10–100 ng dry weight; Gaylor and Hale, unpublished). Again, the majority of these congeners are either absent from technical Aroclor mixtures, or present at less than 0.05% by weight. The major consideration should be that the surrogate compounds represent the range in physical and chemical properties of the analytes of interest.

3. This SFE method is successful using 2 mL GC vials. However, for heavily contaminated samples (>100 μg) it may be necessary to employ larger collection vials. The Accutrap™ unit will accept 12 mL HPLC vials with Teflon-lined septa screw caps. This will facilitate larger desorption volumes if needed. Use of these vials also eliminates the need for any solvent rinsing of the trap between samples. A 1.8-mL solvent rinse between samples is generally performed as a precautionary measure when using 2 mL GC vials for analyte desorption. The Zymark Turbo Vap LV solvent evaporator was designed to accept 15 mL centrifuge vials. The unit was modified to permit sample concentration under a gentle stream of nitrogen directly in the GC/HPLC vials after SFE.
4. Considerable flexibility exists here for the analyst. Any data system capable of analog to digital signal conversion with subsequent peak area integration and quantitation should be adequate. For this work, GC/NCI-MS was the principle analyte-confirmation technique. Numerous studies have shown the applicability of GC-MS (ion trap, SIM, and EI) to analytical SFE as well *(13,25,26)*.
5. A glass fillet board is recommended for use during fish dissection because it is inert, easy to clean, and will withstand rinsing with organic solvents. The fillet board and knife should be rinsed thoroughly with deionized water and the solvent regime described in **Subheading 3., step 1** between samples.
6. The time required for complete drying of tissue samples will vary depending upon sample amount, density, water content and freeze-dryer efficiency. Samples should be checked at 12- to 24-h intervals by probing with clean spatulas. Drying is complete in less than 48 h in most cases. Attempts to dry tissue samples with chemical desiccants during SFE method development failed. It was possible to obtain a sample with a manageable powderlike consistency that appeared visually dry. But, when subjected to SFE, water was released from the sample and often plugged the restrictor and/or trap, ultimately appearing in the final solvent extract. This could be due in part to the elevated temperatures at which the extractions were conducted. Algaier et al. *(27)* reported that raising the extraction temperature from 25°C to 150°C released increasing amounts of water from cotton plugs during SFE using unmodified CO_2. In light of these data, new studies are being conducted in this laboratory to ascertain whether PCBs can be extracted from aquatic biota samples at lower temperatures (higher fluid density) without coextracting sample water. A method has been developed by Capangpangan et al. *(28)* to dry filtered suspended solids from natural water samples before SFE. The technique has been modified to allow drying of small quantities (1–2 g) of wet biota (Hale and Gaylor, unpublished). Wet samples are applied to a glass fiber filter and suspended over a bed of calcium chloride in a closed glass container for 24 h. Assuming successful extraction of a wet sample, any water present in the extract must be removed before GC.
7. Any blender made of glass should suffice for this step. During homogenization of larger fillets, it may be necessary to stop periodically and break up large chunks of tissue with a clean spatula until a powderlike sample consistency is achieved. Solvent rinsed mason jars are excellent for sample storage prior to SFE and long-term archiving.

8. Pentachlorobenzene, PCB 204 and PCB 207 are recommended for internal standards since they are not encountered and do not coelute with sample PCBs during GC separation. Surrogate and internal standards can be prepared from neat or by serial dilution of commercially prepared standards. All standards should be prepared in clean, solvent-rinsed volumetric glassware and stored in a freezer when not in use.
9. Indelible markers are required to label stainless steel extraction vessels because of the high SFE temperatures (150°C) to which they are exposed. Tape is adequate for properly labeling standards, collection vials, and so on. It is essential that working surfaces (i.e., laboratory bench, balance, etc.) be clean during handling, weighing, and loading of extraction vessels to minimize the potential for sample contamination. Failure to remove any excess material from the top rim of the vessel can lead to vessel pressurization problems, resulting in instrument error messages and system shutdown during SFE.
10. Surrogate standards should be formulated in concentrations high enough to minimize the volume of carrier solvent spiked on to the sample before SFE (100 µL recommended). Addition of large solvent volumes can lead to leaching of analytes and subsequent loss through the bottom of the extraction vessel. Further, any solvent remaining in the vessel during SFE can alter the critical parameters of the extraction solvent leading to lipid coextraction and/or reduced analyte extraction efficiency ***(29,30)***.
11. Transfer of the alumina to a 250- to 500-mL beaker after activation is a matter of convenience. The beaker permits the alumina to be poured directly into the vessel without the need for a spatula, thus minimizing the potential for contamination.
12. Trap configuration will vary widely among instruments. The Isco-Suprex trap consists of a stainless steel cylinder with an internal volume of approximately 1.5 mL. This cylinder contains the trapping sorbent. The trap is equipped with two end caps fitted with 1/4 inch internal threads (*see* **Fig. 2**). The top cap is stainless steel and the bottom cap is composed of PEEK. After the trap cylinder is packed and capped on both ends, the trap is connected to a heated, automatic variable restrictor (AVR^{TM}) block via 1/8-inch stainless steel tubing. If Unibeads are unavailable, a 3:1 (w/w) mixture of C_{18}-modified silica and silanized glass beads may be substituted in the cryogenic trap. This combination of materials has shown good retentive capacity for PCBs during SFE. "Fines" removed during the initial trap rinse will be evident by the milk-white color they impart to the rinse solvent. If the solvent is excessively discolored, rinse a second time before proceeding with sample extraction.
13. If it is suspected that a sample is heavily contaminated (>100 µg of PCB, dry weight), 12 mL HPLC vials with Teflon-lined septa screw caps may be used to allow larger desorption volumes. Between 6 and 10 mL of isooctane have proven effective in this laboratory when needed. It is likely, however, that 2-mL vials will be adequate for the majority of applications. Nitrogen gas is purged through the trapping system for approximately 10 seconds after desorption. Nitrogen tanks used for purging solvent and analytes after desorption must be calibrated with a

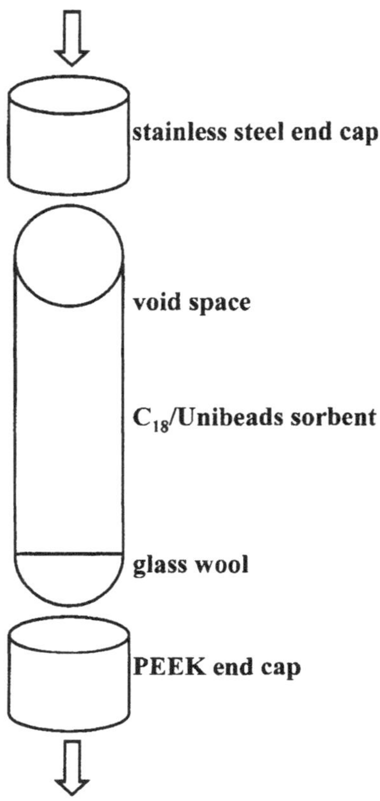

Fig. 2. Diagram of the Accutrap solid phase trap cartridge showing the orientation of C_{18}/Unibeads sorbent, glass wool plug and direction of decompressed CO_2 and desorb solvent flow.

head pressure of 0.4 MPa (~4 atm) when using C_{18} as a trapping sorbent. The AP44™ also requires a constant 0.7 MPa (~7 atm) nitrogen head pressure to actuate pneumatic valves throughout the instrument. It is therefore useful to use a step-down gas regulator to allow a single nitrogen tank to distribute the appropriate head pressure for each function. If it is practical, separate nitrogen tanks can be used for the AP44™ and Accutrap™ units. It is important to note that there are significant differences in design and configuration among the major commercial SFE instruments. It is, therefore, reasonable to assume that differences in extraction efficiency may occur under the same set of extraction conditions between different commercial and "lab-fabricated" instruments ***(31)***. However, there have been reports of attempts to translate SFE methods developed on one instrument to other designs *(32)*. It is recommended that the analyst attempting to translate the SFE protocol described here, regardless of instrument design, begin by conducting test extractions using blank matrices (i.e., alumina, sand, etc.) with amended target analytes. SFE of a previously characterized "real-world" matrix and a Certified Reference Material (CRM) should be conducted for final validation. This SFE method validation approach has been prescribed by other researchers working in analytical SFE *(33–35)*.

14. As with surrogate standard(s) preparation described in **Note 10**, internal quantitation standards should be sufficiently concentrated so as to minimize the spiking solvent volume required (100 µL recommended). This will negate the need for a second solvent reduction step after addition of the internal standard. Repeated solvent reduction can lead to significant analyte losses and subsequent quantitation errors. Again, 12-mL vials can be used to simplify this step. Use of these vials, however, requires that the sample extract be transferred to a GC vial with a pasteur pipette after initial solvent reduction, adding an additional postextraction sample manipulation step to this simple SFE protocol.

Acknowledgments

We thank the Maryland Power Plant Research Program for supporting development of this work under contract No. CB95-002-004. This is contribution number 2289 from the Virginia Institute of Marine Science.

References

1. Eisenberg, M., Mallman, R., and Tubiash, H. (1980) Polychlorinated biphenyls in fish and shellfish of the Chesapeake Bay. *Marine Fish. Rev.* **42,** 21–25.
2. McFarland, V. A. and Clarke, J. U. (1989) Environmental occurrence, abundance and potential toxicity of polychlorinated biphenyl congeners: considerations for a congener-specific analysis. *Environ. Health Perspect.* **81,** 225–239.
3. Subramanian, B. R., Tanabe, S., Hidaka, H., and Tatsukawa, R. (1983) DDTs and PCB isomers and congeners in Antarctic fish. *Arch. Environ. Contam. Toxicol.* **12,** 621–626.
4. Rubinstein, N. I, Gilliam, W. T., and Gregory, N. R. (1984) Dietary accumulation of PCBs from a contaminated source by a demersal fish (*Leiostomus Xanthrus*). *Aquat. Toxicol.* **5,** 331–342.
5. Schneider, R. (1982) Polychlorinated biphenyls (PCBs) in cod tissues from the Western Baltic: significance of equilibrium partitioning and lipid composition in the bioaccumulation of lipophilic pollutants in gill-breathing animals. *Sounderdruck Bd.* **29,** 69–79.
6. Clark, J. R., Patrick, J. M., Moore, J. C., and Forester, J. (1986) Accumulation of sediment-bound PCBs by fiddler crabs. *Bull. Environ. Contam. Toxicol.* **36,** 571–578.
7. Hale, R. C. and Smith, C. L. (1996) A multiresidue approach for trace organic pollutants: application to effluents and associated aquatic sediments and biota from the southern Chesapeake Bay drainage basin 1985–1992. *Int. J. Environ. Anal. Chem.* **64,** 21–33.

8. Long, A. R., Soliman, M. M., and Barker, S. A. (1991) Matrix solid phase dispersion (MSPD) extraction and gas chromatographic screening of nine chlorinated pesticides in beef fat. *J. Assoc. Offic. Anal. Chem.* **74,** 493.
9. Van der valk, F. and Wester, P. G. (1991) Determination of toxaphene in fish from Northern Europe. *Chemosphere* **22,** 57.
10. Hale, R. C. and Greaves, J. (1992) Methods for the analysis of persistent chlorinated hydrocarbons in tissues. *J. Chromatogr.* **580,** 257–278.
11. Hale, R. C. and Gaylor, M. O. (1995) Determination of PCBs in fish tissues using supercritical fluid extraction. *Environ. Sci. Technol.* **29,** 1043–1047.
12. Camel, V., Tambuté, A., and Caude, M. (1993) Analytical-scale supercritical fluid extraction: a promising technique for the determination of pollutants in environmental matrices. *J. Chromatogr.* **642,** 263–281.
13. Bøwadt, S. and Johansson, B., Wunderli, S., Zennegg, M., de Alencastro, L. F., and Grandjean, D. (1995) Independent comparison of Soxhlet and supercritical fluid extraction for the determination of PCBs in an industrial soil. *Anal. Chem.* **67,** 2424–2430.
14. Bøwadt, S. and Johansson, B. (1994) Analysis of PCBs in sulfur-containing sediments by off-line supercritical fluid extraction and HRGC-ECD. *Anal. Chem.* **66,** 667–673.
15. Onuska, F. I., Terry, K. A., and Wilkinson, R. J. (1993) The analysis of chlorinated dibenzofurans in municipal fly ash: supercritical fluid extraction vs Soxhlet. *J. High Resol. Chromatogr.* **16,** 407–412.
16. Hale, R. C. and Gaylor, M. O. (1996) Robustness of supercritical fluid extraction (SFE) in environmental studies: analysis of chlorinated pollutants in tissues from the osprey (*Pandion haliaetus*) and several fish species. *Int. J. Environ. Anal. Chem.* **64,** 11–19.
17. Johansen, H. R., Becher, G., and Greibrokk, T. (1992) Determination of PCBs in biological samples using on-line SFE-GC. *Fresenius J. Anal. Chem.* **344,** 486–491.
18. Taylor, L. T. (1996) *Supercritical Fluid Extraction.* Wiley, New York, pp. 136–138.
19. Hawthorne, S. B., Langenfeld, J. J., Miller D. J., and Burford, M. D. (1992) Comparison of supercritical $CHClF_2$, N_2O and CO_2 for the extraction of polychlorinated biphenyls and polycyclic aromatic hydrocarbons. *Anal. Chem.* **64,** 1614–1622.
20. Crowther, J. B. and Henion, J. D. (1985) Supercritical fluid chromatography of polar drugs using small-particle packed columns with mass spectrometric detection. *Anal. Chem.* **57,** 2711–2716.
21. Burford, M. D., Hawthorne, S. B., and Miller, D. J. (1993) Evaluation of drying agents for off-line supercritical fluid extraction. *J. Chromatogr. A* **657,** 413–427.
22. Greaves, J., Harvey, E., and Huggett, R. J. (1991) Evaluation of gas chromatography with electrolytic conductivity detection and electron capture detection and use of negative chemical ionization GC-MS for the analysis of PCBs in effluents. *Environ. Toxicol. Chem.* **10,** 1391–1398.
23. Analytical Protocol for Hazardous Organic Chemicals in Environmental Samples. (1991) Division of Chemistry and Toxicology, Virginia Institute of Marine Science, School of Marine Science, College of William and Mary. Special Publication REFSH001.V48 (131) 68p.

24. Schulz, D. E., Petrick, G., and Duinker, J. C. (1989) Complete characterization of polychlorinated biphenyl congeners in commercial aroclor and clophen mixtures by multidimensional gas chromatography-electron capture detection. *Environ. Sci. Technol.* **23,** 852–859.
25. Johansen, H. R., Becher, G., and Greibrokk, T. (1994) Determination of planar PCBs by combining on-line SFE-HPLC and GC-ECD or GC/MS. *Anal. Chem.* **66,** 4068–4073.
26. Supercritical Fluid Extraction of Environmental Pollutants from Animal Tissues. (1993) *Application Note 310,* Publication #LPN034884 Dionex Corporation, Atlanta, GA.
27. Algaier, J. and Tehrani, J. (1993) The effect of selected sorbents on water management trapping in SFE. Presented at the Pittsburgh Conference (PITTCON '93), Paper #395, March.
28. Capangpangan, M. B. and Suffet, I. H. (1996) Optimization of a drying method for filtered suspended solids from natural waters for supercritical fluid extraction analysis of hydrophobic organic compounds. *J. Chromatogr. A* **738,** 253–264.
29. Hawthorne, S. B., Miller, D. J., Burford, M. D., Langenfeld, J. J., Eckert-Tilotta, S., and Louie, P. K. (1993) Factors controlling quantitative supercritical fluid extraction of environmental samples. *J. Chromatogr.* **642,** 301–317.
30. Järvenpää, E., Huopalahti, R., and Tapanainen, P. (1996) Use of supercritical fluid extraction-high performance liquid chromatography in the determination of polynuclear aromatic hydrocarbons from smoked and broiled fish. *J. Liquid Chromatogr. Relat. Technol.* **19,** 1473–1482.
31. Lopez-Avila, V., Dodhiwala, N. S., Benedicto, J., and Beckert, W. F. (1991) Evaluation of four supercritical fluid extraction systems for extracting organics from environmental samples. *LC-GC* **10,** 762–769.
32. King, J. W., Snyder, J. M., Taylor, S. L., Johnson, J. H., and Rowe, L. D. (1993) Translation and optimization of supercritical fluid extraction methods to commercial instrumentation. *J. Chromatogr. Sci.* **31,** 1–5.
33. Engelhardt, H., Zapp, J., and Kolla, P. (1991) Sample preparation by supercritical fluid extraction in environmental, food, and polymer analysis. *Chromatographia* **32,** 527–537.
34. Kuitunen, M. L., Hartonen, K., and Riekkola, M. L. (1991) Analysis of chemical warfare agents in soil samples by off-line supercritical fluid extraction and capillary gas chromatography. *J. Microcolumn Sep.* **3,** 505–512.
35. Benner, B. A. (1993) Standard reference materials for use in supercritical fluid extraction method development. Presented before the Division of Environmental Chemistry. *Proc. Am. Chem. Soc.* **33,** 324–326.

6

Isolation of Polynuclear Aromatic Hydrocarbons from Fish Products by Supercritical Fluid Extraction

Eila P. Järvenpää and Rainer Huopalahti

1. Introduction

Polynuclear or polycyclic aromatic hydrocarbons (PAHs) are mutagenic compounds formed by incomplete burning of organic material. The mutagenity and carcinogenic activity becomes higher as the number of fused rings in a molecule increases *(1)*. The human intake of PAHs is very variable. The main sources are industrial and automobile exhaust gases and tobacco smoke. A percentage of the intake is obtained from baked, smoked, and grilled foodstuffs. This food-originating portion depends upon the habits of food consumption, the foodstuffs themselves, and the manufacturing methods *(1–4)*.

Usually, solvent extraction methods using chlorinated solvents followed by solid-phase extraction clean-up and chromatographic determination are needed to analyze the PAH content of foods *(2–4)*. To some extent, the use of supercritical fluid extraction or SFE (*see* Chapter 1) instead of liquid solvent extraction has decreased the number of clean-up steps needed *(5)*. SFE has already been used for the determination of PAHs from environmental samples *(5–9)*. In soil and food samples, the factor limiting the extractability of PAHs is matrix interactions, not solubility. It has been shown that these compounds bind very strongly to the matrix components and some difficulties with SFE may occur *(6–9)*. These problems have been overcome by increasing the extraction temperature and the solvating power *(6–8)*. The latter is accomplished by adding cosolvents (modifiers) to supercritical fluids *(8)*.

This chapter describes an SFE method that can be used to isolate PAHs from fish tissues. Quantitation is by high performance liquid chromatography (HPLC). The protocol can be used (with minor modifications) to estimate the

From: *Methods in Biotechnology, Vol. 13: Supercritical Fluid Methods and Protocols*
Edited by: J. R. Williams and A. A. Clifford © Humana Press Inc., Totowa, NJ

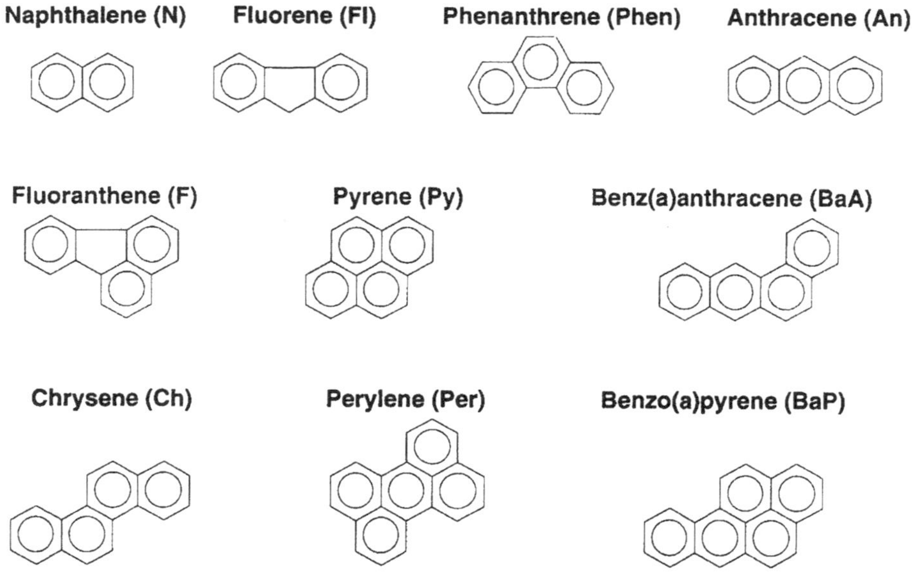

Fig. 1. The structures of the PAHs determined in this study.

amount of PAHs in different foods. However, the method was developed with dehydrated fish samples *(10)*.

2. Materials

1. Liquid and solid chemicals: all solvents (methanol, acetonitrile, water, dichloromethane) should be HPLC-grade. Adsorbents (silica gel 60, aluminum oxide 90) and quartz sand can be used directly from their containers.
2. Carbon dioxide: grade 4.8 or SFE-grade with helium head pressure.
3. Reference compounds: pure PAHs (E. Merck, Darmstadt, Germany; Sigma Chem., St. Louis, MO; or equivalent) are needed for quantitative determination because each of them has a different response by ultraviolet (UV) detection. In practice, the responses of each component are calculated in relation to naphthalene (standard). The structures of the compounds determined in this study are shown in **Fig. 1**.
4. SFE equipment and accessories: an ISCO SFX 220 apparatus (Isco Inc., Lincoln, NE) with two pumps for fluid delivery was used. The addition of cosolvent is necessary for this application (*see* **Note 1**). The flow rate is set with a linear silica capillary restrictor. However, other types of back-pressure regulator can be used as well.
5. HPLC equipment and accessories: a binary solvent delivery system with the possibility of gradient programming is needed. PAHs are detected with a UV detector (wavelength 254 nm) and peak areas are measured with an integrator. It is recommended that standardized injection volume (loop size, e.g., 20 µL) be used for greater repeatability.

3. Method

1. Sample preparation: homogenize the edible parts of the fish samples and lyophilize to low water content (about 1%). Mix the freeze-dried material thoroughly and keep it in an airtight container in a refrigerator. It is not recommended to keep the samples for a long time. A 1-g portion of the freeze-dried fish is weighed accurately. Mix it with 1 g of quartz sand (*see* **Note 2**). Put this combined sample into an extraction vessel of volume 2.5 mL (*see* **Note 3**).
2. SFE: before extraction, keep the filled vessel for 10 min in the extraction chamber (oven) at 70°C to achieve thermal equilibration. The binary fluid system consists of carbon dioxide (CO_2) modified with 10% (v/v) methanol. Extract the samples at 70°C and 350 atm with 20 mL of fluid at a flow rate of about 1.6 mL/min (*see* **Note 4**). Collect the analytes in a 15- to 20-mL test tube containing 3 mL of hexane:dichloromethane (3:1, v/v).
3. Purification of the extracts: prepare clean-up columns using Pasteur pipettes. Prepare columns by measuring 1.0 g of aluminum oxide (column a) and 0.8 g of silica gel 60 (column b) over glass wool plugs. Commercial clean-up columns can be used as well. Elute the analytes through columns (a) and (b) put in series with 2 × 1.5 mL of hexane:dichloromethane (3:1, v/v). Add 2 mL of acetonitrile to the collected eluate and evaporate to 1 mL volume. Elute this solution through a C_{18} cartridge and wash with 2 mL of acetonitrile. Collect all the eluate and add 1 μg of naphthalene as a standard (*see* **Note 5**).
4. Quantitative determination: PAHs are determined by reverse phase HPLC with UV detection at 254 nm as follows. An ODS column (LiChrospher C-18, 250 × 4 mm, 5 μm particle size; or equivalent column) can be used for the separation. The resolution shown in **Fig. 2** is obtained with a gradient of 50 to 98% acetonitrile in water over 24 min at a flow rate of 0.8 mL/min (*see* **Note 6**).

4. Notes

1. If the equipment used does not facilitate modifier addition, methanol (1 mL) could be added directly to the extraction cell before extraction. In this case, proceed with system testing (*see* **Note 3**) carefully to determine the extraction recovery and, if necessary, adjust modifier volume.
2. Quartz sand is used to enhance the fluid flow through the sample and fill the void volume of the extraction vessel.
3. In this work, the extraction vessel volume was 2.5 mL. However, larger sample sizes with bigger vessels can be used.
4. Conditions for extraction system testing: prepare spiked samples. A) Fill the extraction vessel with quartz sand spiked with standard PAHs (e.g., the solution used for determination by HPLC). Extract and analyze the sample as described in **Subheading 3., steps 2–4**, except that the purification steps are not needed. The recoveries should be around 100% [relative standard deviation (RSD) 3–11%]. B) Lyophilized fish tissue (not contaminated) spiked with PAHs is extracted as described in **Subheading 3.** The recoveries should be 80–100% with an RSD of 3–11% (*see* **Fig. 3**). Further discussion can be found in ref. ***10***.

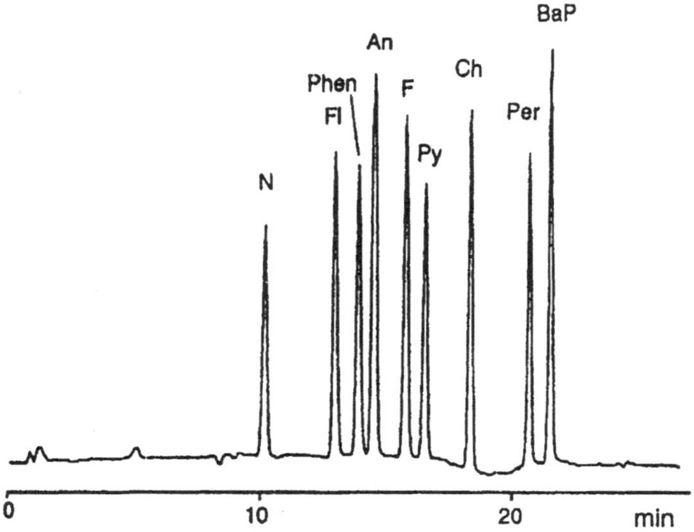

Fig. 2. An example chromatogram of a PAH standard solution. For HPLC conditions, see text. N, naphthalene (std); Fl, fluorene; Phen, phenanthrene; An, anthracene; F, fluoranthene; Py, pyrene; Ch, chrysene; Per, perylene; BaP, benzo(a)pyrene.

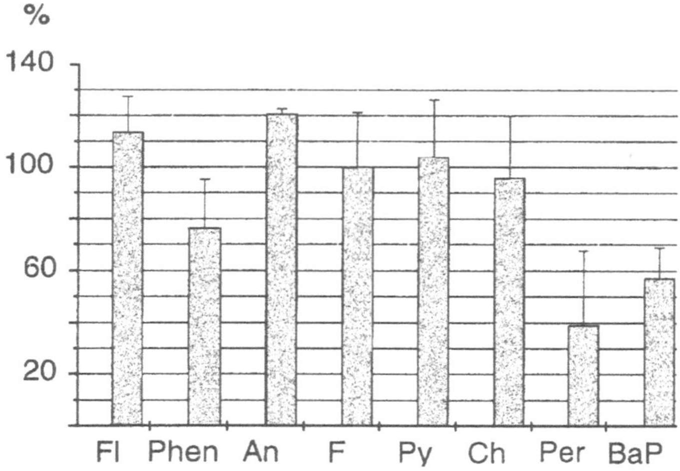

Fig. 3. Recoveries of selected PAHs from spiked fish sample. For abbreviations, see Fig. 1.

5. A constant amount of naphthalene (e.g., 10 µL of standard containing 0.1 µg/µL) is added to the sample solutions in order to determine the exact volume of solution. The peak areas obtained are compared to those of standard chromatograms. This comparison combined with the result obtained from the naphthalene response gives the concentration of the analytes in the sample.

6. Water and acetonitrile produce gas when mixed. Consequently, use pure acetonitrile in one solvent container and water:acetonitrile (50:50, v/v), sonicated and/or degassed, in another solvent container to minimize gas problems in the HPLC system.

References

1. Cooke, M. and Dennis, A. J. (1986) *Polynuclear Aromatic Hydrocarbons: Chemistry, Characterization and Carcinogenesis*. Battelle Press, Columbus, OH.
2. Gomaa E. A., Gray J. I., Rabie S., Lopez-Bote C., and Booren A. M. (1993) Polycyclic aromatic hydrocarbons in smoked food products and commercial smoke flavourings. *Food Addit. Contam.* **10,** 503–521.
3. Joe Jr, F. L., Salemme, J., and Fazio, T. (1984) Liquid chromatographic determination of trace residues of polynuclear aromatic hydrocarbons in smoked foods. *J. AOAC* **67,** 1076–1082.
4. Perfetti, G. A., Nyman, P. J., Fisher, S., Joe Jr, F. L., and Diachenko, G. W. (1992) Determination of polynuclear aromatic hydrocarbons in seafood by liquid chromatography with fluorescence detection. *J. AOAC Int.* **75,** 872–877.
5. Reimer, G. and Suarez, A. (1995) Comparison of supercritical fluid extraction and Soxhlet extraction for the analysis of native polycyclic aromatic hydrocarbons in soils. *J. Chromatogr. A* **699,** 253–263.
6. Reindl, S. and Höfler, F. (1994) Optimization of the parameters in supercritical fluid extraction of polynuclear aromatic hydrocarbons from soil samples. *Anal. Chem.* **66,** 1808–1816.
7. Janda, V., Bartle, K. D., and Clifford, A. A. (1993) Supercritical fluid extraction in environmental analysis. *J. Chromatogr.* **642,** 283–299.
8. Bøwadt, S. and Hawthorne, S. B. (1995) Supercritical fluid extraction in environmental analysis. *J. Chromatogr. A* **703,** 549–571.
9. Monserrate, M. and Olesik, S. V. (1997) Evaluation of SFE-CO_2 and methanol-CO_2 mixtures for the extraction of polynuclear aromatic hydrocarbons from house dust. *J. Chromatogr. Sci.* **35,** 82–90.
10. Järvenpää, E., Huopalahti, R., and Tapanainen, P. (1996) Use of supercritical fluid extraction-high performance liquid chromatography in the determination of polynuclear aromatic hydrocarbons from smoked and broiled fish. *J. Liq. Chromatogr.* **19,** 1473–1487.

7

Supercritical Fluid Extraction of Mycotoxins from Feeds

Rainer Huopalahti and Eila P. Järvenpää

1. Introduction

Trichothecenes are sesquiterpenoid mycotoxins produced by a variety of species of imperfect fungi. These mycotoxins are found mainly as products of field flora in grains and cereals. Trichothecenes show a wide range of toxicity, which is dependent on the structure of the molecule. Over 150 trichothecenes have been isolated and characterized, but it is still a challenging analytical task to isolate and characterize these compounds from foods and feeds *(1,2)*.

Conventional methods for the isolation of trichothecenes involve extensive and time-consuming sample preparation steps. According to a recent survey, two-thirds of analysis time is devoted to sample preparation and this step accounts for at least one-third of the errors generated during the performance of an analytical method *(3)*. Supercritical fluid extraction or SFE (*see* Chapter 1) has shown great potential and can offer shorter extraction times, higher recoveries and lower consumption of organic solvents than with conventional solvent extraction.

Mycotoxins are quite often separated, identified, and quantitated using thin-layer chromatography *(4)*, thin-layer chromatography/mass spectrometry *(5)*, gas chromatography-mass spectrometry (GC-MS) *(6–10)*, and high-performance liquid chromatography (HPLC) methods with fluorescence *(11,12)* or light-scattering *(13)* detection. Conventional HPLC and GC methods, however, suffer serious drawbacks. The sensitivity of HPLC is limited, since most trichothecenes have minimal fluorescent or ultraviolet-absorbing properties. In the case of GC methods, derivatization is often required, which may cause problems with quantitative analysis procedures. HPLC combined with mass

From: *Methods in Biotechnology, Vol. 13: Supercritical Fluid Methods and Protocols*
Edited by: J. R. Williams and A. A. Clifford © Humana Press Inc., Totowa, NJ

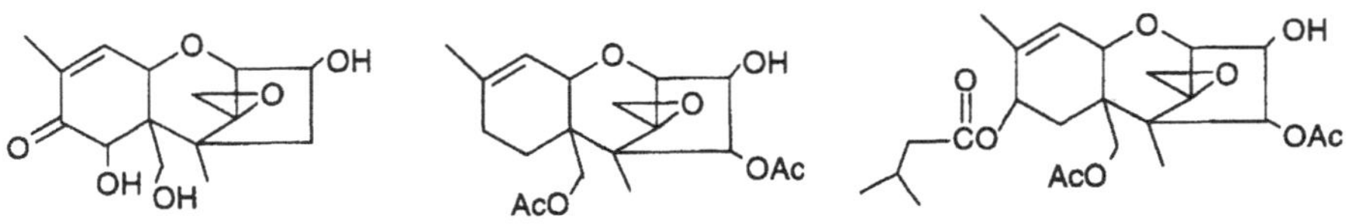

DON DAS T-2 toxin

Fig. 1. The structures of the three trichothecenes investigated.

spectrometry via thermospray, plasmaspray/ionspray or fast atom bombardment *(14–22)* has also been reported.

This chapter describes an SFE method for the isolation of trichothecene mycotoxins from grain-based feeds. Quantitative or near-quantitative recoveries of 4-deoxynivalenol (4-DON), diacetoxyscirpenol (DAS), and T-2 toxin (T-2) (*see* **Fig. 1**) are possible using supercritical carbon dioxide–methanol as the extraction fluid. In this study, quantitation was made by HPLC combined with ionspray mass spectrometry. Alternative quantitation methods can be used as well, for example, UV-detection and enzyme immunoassay techniques for the determination of 4-DON in supercritical fluid extracts of grain samples are described in ref. ***23***.

2. Materials

1. Ground samples to small particles of uniform size using a mill or a homogenisator (e.g., Moulinette S food processor, Moulinex, France; a coffee grinder or equivalent depending upon the sample type). Store ground samples in capped plastic containers at room temperature.
2. Reference compounds: a stock solution (500 ng/µL) of the three mycotoxins, 4-DON, DAS and T-2 (Sigma Chemicals, St. Louis, MO), is prepared in HPLC-grade methanol. In the experiments described below, the stock solution was diluted 1:100 with methanol.
3. Carbon dioxide: SFC-grade CO_2 with helium head pressure in a cylinder equipped with a diptube.
4. Organic solvents should be preferably HPLC-grade. The purity of other liquid and solid chemicals should be at least reagent-grade.
5. SFE equipment and accessories: the addition of modifier to supercritical carbon dioxide is necessary in this application. For example, an ISCO Model 100 DX dual syringe pump system coupled with an ISCO SFX 2-10 extractor (Isco Inc., Lincoln, NE) and a two channel adjustable restrictor device can be used. The head of the carbon dioxide pump is maintained at 5°C with an external cryostat. The other pump of the ISCO system is used for adding methanol modifier dynamically (*see* **Note 1**).

6. HPLC equipment capable of low flow rates is needed for the analyses, e.g., the Waters Model 600-MS system (Waters Inc., Milford, MA) can be used. The column used in this protocol in conjunction with mass spectrometry is Betasil C18 (100 mm × 2 mm ID, 5 µm particles, 100 Å) (Keystone Scientific, Bellafonte, PA), but equivalent columns can be used as well. The trichothecenes can be monitored using a UV-detector at 195–225 nm (*see* **Note 2**). In this application, more specific detection is achieved using mass spectrometry.
7. Ionspray LC/MS, for mass spectrometric detection, for example, the PE Sciex API 300 LC/MS/MS system (Perkin Elmer, Thornhill, ON). The operation conditions are described in **Subheading 3., step 5**.

3. Method

1. Preparation of fortified samples: spiked samples are used for testing the performance of the SFE system. Inject an appropriate amount of the diluted standard solution of mycotoxins on to the noncontaminated sample in a 10 mL extraction vessel. The concentrations used are 250, 500, and 1500-ppb. For recovery requirements, *see* **Note 3**. Repeatability of SFE is based on the data obtained from the tests, where samples were spiked before and after SFE.
2. SFE: weigh 4 g of sample into the extraction vessel (10 mL), then fill the void volume with anhydrous sodium sulfate. Equilibrate the sample in the extraction chamber at the extraction temperature for 10 minutes. Use the following extraction conditions: fluid composition 5% (v/v) methanol in carbon dioxide, pressure 550 atm (1 atm = 0.10132 MPa), temperature 60°C, restrictor temperature 65°C, and fluid volume 30 mL. Set the flow rate of supercritical fluid at about 1.2 mL/min. Collect the analytes by bubbling the extracted material into 10 mL of methanol in a test tube of volume 20–25 mL.
3. Preparation of the samples after SFE: remove fat from the SFE-derived extracts with 3 × 2 mL of hexane. Discard the hexane layers. Evaporate the residual solvent(s) with nitrogen, and dissolve with 500 µL of HPLC mobile phase (*see* **Subheading 3., step 4**) and store at +6°C prior to quantitative determination.
4. The extracts are analyzed by HPLC using an ODS reversed phase column. The mobile phase consists of methanol, acetonitrile and aqueous ammonium acetate (3 m*M*) (45:5:50, v/v/v). A suitable flow rate for the above mentioned column used with mass spectrometry is 0.2 mL/min. The elution order is DON, DAS, T-2 toxin, and the retention times are verified using the standard solution.
5. In this application, quantitation is made by measuring ammonium adduct ions produced in the ionspray interface of the LC/MS system. The mycotoxins can be determined under full-scan and selected-ion monitoring modes. The mass range in full-scan experiments is m/z 50–600, scan rate 4 s/scan. For better selectivity, selected ion monitoring can be used. The ions monitored in this case are $[M + H]^+$ and $[M + NH_4]^+$, i.e., 297 and 314.2 for DON, 367 and 384.1 for DAS, and 467 and 484.1 for T-2 toxin, respectively. The quantitation in both methods is based on the responses obtained using the reference solutions described in **Subheading 2.** For further details of this method *see* **ref. *22***.

4. Notes

1. If the SFE equipment available does not facilitate modifier addition, methanol (1 mL) could be added directly to the extraction cell over the sample before extraction. In this case, proceed with system testing (*see* **Note 3**) carefully to ascertain the recovery by SFE, and if needed, adjust the volume of methanol added.
2. Further clean-up of sample extracts is often needed if UV-detection is used in quantitation. For most of the sample types, simple Florisil clean-up columns prepared in Pasteur pipettes are sufficient. Procedure: add 2 g of Florisil in hexane to a Pasteur pipette over a glass wool plug and a layer of anhydrous sodium sulfate. Add some sodium sulfate on to the top. Add the sample in methanol (about 0.5 mL) and wash the column with 20 mL of hexane. Elute the analytes using 25 mL of chloroform:methanol (9:1, v/v). Evaporate the eluate and dissolve the residue with HPLC mobile phase.
3. Conditions for extraction system testing: prepare spiked samples as described in **Subheading 3., step 1**. Extract and analyze the fortified samples. The recoveries obtained should be about 95% for DON and 85% for DAS and T-2 toxin. If recoveries are inadequate, increase the fluid volume.

References

1. Ueno, Y. (1983) *Trichothecenes: Chemical, Biological and Toxicological Aspects.* Elsevier, Amsterdam.
2. Betina, V. (1989) *Mycotoxins: Chemical, Biological and Environmental Aspects.* Elsevier, Amsterdam.
3. Majors, R. E. (1991) An overview of sample preparation. *LC-GC Int.* **4,** 10–14.
4. Sano, A., Asabe, Y., Takitani, S., and Ueno, Y. (1982) Fluorodensitometric determination of trichothecene mycotoxins with nicotinamide and 2-acetylpyridine on a silica gel layer. *J. Chromatogr.* **235,** 257–265.
5. Tripathi, D. N., Chauhan, L. R., and Bhattacharya, A. (1991) Separation and identification of mycotoxins by thin-layer chromatography/fast atom bombardment mass spectrometry. *Anal. Sci.* **7,** 423–435.
6. Black, R. M., Clarke, R. J., and Read, R. W. (1987) Detection of trace levels of trichothecene mycotoxins in environmental residues and foodstuffs using gas chromatography with mass spectrometric or electron-capture detection. *J. Chromatogr.* **388,** 365–378.
7. Kostiainen, R. and Rizzo, A. (1988) The characterization of trichothecenes as their heptafluorobutyrate esters by negative-ion chemical ionization tandem mass spectrometry. *Anal. Chim. Acta* **204,** 233–246.
8. Plattner, R. D., Beremand, M. N., and Powell, R. G. (1989) Analysis of trichothecene mycotoxins by mass spectrometry and tandem mass spectrometry. *Tetrahedron* **45,** 2251–2262.
9. Kostiainen, R. and Nokelainen, S. (1990) Use of M-series retention index standards in the identification of trichothecenes by electron impact mass spectrometry. *J. Chromatogr.* **513,** 31–37.

10. Schwadorf, K. and Müller, H.-M. (1991) Determination of trichothecenes in cereals by gas chromatography with ion trap detection. *Chromatographia* **32,** 137–142.
11. Kok, W. T. (1994) Derivatization reactions for the determination of aflatoxins by liquid chromatography with fluorescence detection. *J. Chromatogr. B* **659,** 127–137.
12. Shephard, G. S., Thiel, P. G., and Sydenham, E. W. (1995) Liquid chromatographic determination of the mycotoxin fumonisin B_2 in physiological samples. *J. Chromatogr.* **692,** 39–43.
13. Wilkes, J. G., Sutherland, J. B., Churchwell, M. I., and Williams, A. J. (1995) Determination of fumonisins B1, B2, B3 and B4 by high-performance liquid chromatography with evaporative light-scattering detection. *J. Chromatogr.* **695,** 319–323.
14. Voyksner, R. D., Hagler Jr., W. M., and Swanson, S. P. (1987) Analysis of some metabolites of T-2 toxin, diacetoxyscirpenol and deoxynivalenol by thermospray high-performance liquid chromatography-mass spectrometry. *J. Chromatogr.* **394,** 183–199.
15. Rajakylä, E., Laasasenaho, K., and Sakkers, P. J. D. (1987) Determination of mycotoxins in grain by high-performance liquid chromatography and thermospray liquid chromatography-mass spectrometry. *J. Chromatogr.* **384,** 391–402.
16. Kostiainen, R. (1991) Identification of trichothecenes by thermospray, plasmaspray and dynamic fast-atom bombardment liquid chromatography-mass spectrometry. *J. Chromatogr.* **562,** 555–562.
17. Holcomb, M., Sutherland, J. B., Chiarelli, M. P., Korfmacher, W. A., Thompson Jr., H. C., Lay Jr., J. O., Hankins, J. L., and Cerniglia, C. E. (1993) HPLC and FAB mass spectrometry analysis of fumonisins B_1 and B_2 produced by *Fusarium moniliforme* on food substrates. *J. Agric. Food Chem.* **41,** 357–360.
18. Young, J. C. and Games, D. E. (1993) Analysis of *Fusarium* mycotoxins by supercritical fluid chromatography with ultraviolet or mass spectrometric detection. *J. Chromatogr.* **653,** 372–379.
19. Kalinoski, H. T., Udseth, H. R., Wright, B. W., and Smith, R. D. (1988) Supercritical fluid extraction and direct fluid injection mass spectrometry for the determination of trichothecene mycotoxins in wheat samples. *Anal. Chem.* **58,** 2421–2425.
20. Taylor, S. L., King, J. W., Richard, J. L., and Greer, J. I. (1993) Analytical-scale supercritical fluid extraction of aflatoxin B_1 from field-inoculated corn. *J. Agric. Food Chem.* **41,** 910–913.
21. Engelhardt, H. and Haas, P. (1993) Possibilities and limitations of SFE in the extraction of aflatoxin B1 from food matrices. *J. Chromatogr. Sci.* **31,** 13–19.
22. Huopalahti, P. R., Ebel Jr., J., and Henion, J. D. (1997) Supercritical fluid extraction of mycotoxins from feeds with analysis by LC/UV and LC/MS. *J. Liq. Chromatogr. Relat. Technol.* **20,** 537–551.
23. Järvenpää, E. P., Taylor, S. L., King, J. W., and Huopalahti, R. (1997) The use of supercritical fluid extraction for the determination of 4-deoxynivalenol in grains: the effect of the sample clean-up and analytical methods on quantitative results. *Chromatographia* **46,** 33–39.

8

Supercritical Fluid Extraction of Pigments from Seeds of *Eschscholtzia californica* Cham

Maria L. Colombo and Andrea Mossa

1. Introduction

Eschscholtzia californica Cham. or the California poppy is the state flower of California. The chemical constituents of the epigeous parts of this plant have been extensively investigated for their isoquinoline alkaloid components *(1,2)*. The hydroalcoholic tincture of the blooming aerial parts is used as an analgesic and sedative even if it does not contain morphinane alkaloids *(3)*. Few reports are known about *E. californica* seeds and their phytochemical pattern is poorly studied *(4–8)*.

As a first step in our study, we examined the *E. californica* seed germination correlated with the turnover of the main secondary metabolites (red pigments) extracted with organic solvents *(6,7)*. We found that the colored components of *E. californica* seeds are lipophilic compounds soluble in *n*-hexane at room temperature. Then, we carried out the extraction of these compounds (red pigments) with supercritical carbon dioxide *(9)*.

The purpose of this chapter is to present a simple and effective protocol for the extraction of red pigments from *E. californica* seeds with supercritical carbon dioxide (CO_2). Supercritical fluid extraction (SFE) was introduced in Chapter 1.

2. Materials

1. Seeds of *E. californica* are commercial seeds purchased from a local market, F. lli Ingegnoli, Milano, Italy (*see* **Note 1**). The seeds are finely ground in a blender to produce particles of 1 to 1.5 mm in diameter (*see* **Note 1**).
2. The SFE unit is a laboratory scale plant (Fedegari Autoclavi spa, Albuzzano, Pavia, Italy) designed to treat solids with supercritical CO_2. SFE is performed with the total recycling of carbon dioxide used as the solvent.

From: *Methods in Biotechnology, Vol. 13: Supercritical Fluid Methods and Protocols*
Edited by: J. R. Williams and A. A. Clifford © Humana Press Inc., Totowa, NJ

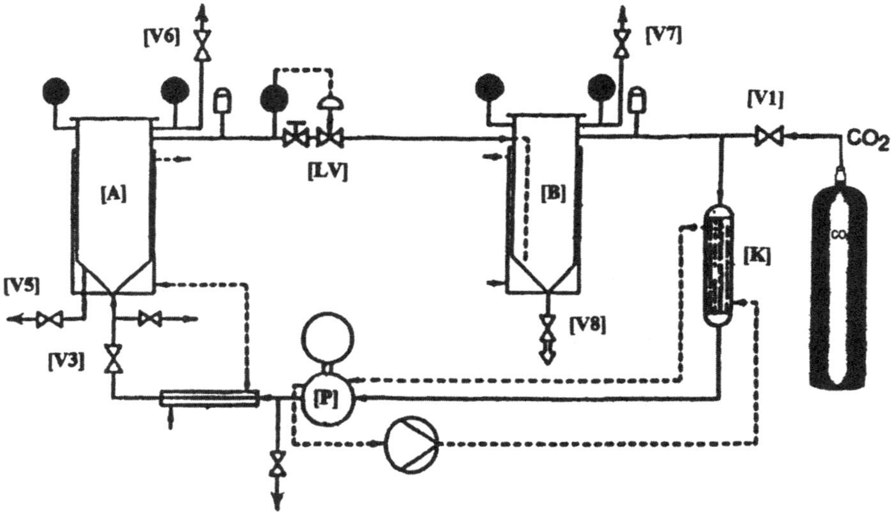

Fig. 1. Schematic diagram of the supercritical CO_2 off-line extractor unit.

3. The basic components of the off-line extractor (*see* **Fig. 1**) are
   a. 30-kg bottle of commercial CO_2 (purity 99%) at 3.94 to 4.93 MPa (40 to 50 atm) pressure, with a diptube;
   b. condenser [K], the CO_2 liquifier, for the condensation of gaseous CO_2, with an internal heat exchanger, refrigerated by a Freon compressor;
   c. metering piston pump [P] for liquid CO_2;
   d. high-pressure needle valves, for feed and recycling of CO_2;
   e. stainless steel extraction autoclave, the extractor [A] of 350 cm^3 external volume, equipped inside with a cylindrical basket (200 cm^3) fitted with sintered metal filters on both ends, which contains the ground solid to be treated, the cylinder basket outside has a Teflon guard O-ring to ensure the pressure seal; the extractor has a screw lid fitted with a device that avoids opening of the lid until there is no pressure in the extractor; the extractor has a safety valve set to open at 54.24 MPa (about 500 bar);
   f. stainless steel separator, the extract accumulator or trap or collector [B], of 350 cm^3 internal volume fitted with two lateral quartz windows; the separator has a safety valve set to open at 7.89 MPa (about 80 bar);
   g. laminating valve [LV] between the extractor and the separator is a pressure controller valve between the extractor and the separator. This is an on/off pneumatic valve which is served by gaseous nitrogen or compressed air at 0.5 MPa (about 5 atm);
   h. each of the two autoclaves (extractor and separator) is provided with separate temperature control (by the circulation of warm water) from 20 to 80°C.

3. Method

3.1. SFE

3.1.1. Starting Procedure

1. Select the pressure on the electrical board. To obtain a constant temperature for extraction (40°C) , set the warm water bath about 5°C higher.
2. Open the nitrogen bottle.
3. Load into the cylindrical basket 50 g of ground raw material for each extraction, gently tapping until the cylindrical basket is filled (do not press the ground material). Close the basket with the sintered upper filter and insert the Seeger ring.
4. Switch on the CO_2 liquifier (condenser).
5. Open the carbon dioxide bottle: liquid CO_2 flows through the [V1] valve to the condenser. The pressure values are kept between 3.94 and 4.93 MPa (about 40 to 50 bar) by a manostat switching a compressor on and off. Close the CO_2 bottle.
6. The liquid CO_2 reaches the pump. The pump-head is cooled. The rate of the pump is regulated by a screw, which controls travel of the ram. The operating flow rate of CO_2 is 3 kg/h or 0.83 ± 0.01 g/s.
7. Open valve [V3] in order to permit CO_2 (from the pump) to reach the extractor.
8. Open valves [V6] and [V7] to permit the exit of air from the rig.
9. A short time (4 to 5 min) later, the rig is filled with CO_2.
10. Close valves [V6] and [V7], and start operation of the pump.

3.1.2. Extraction Procedure

1. The pump increases the pressure of the liquid CO_2 so that it is above its critical pressure of 7.18 MPa or 72.9 atm/bar (*see* **Note 2**). The pumped CO_2 flows through the double-jacketed heated coil and enters the extractor from the bottom through the ground matrix.
2. Always check the laminating valve to see that it is working properly. It must be opened when the desired pressure is reached and then closed. This device is important and it is to be checked during the entire extraction time.
3. The dissolved compounds, extracted by supercritical CO_2 from the matrix, arrive in the separator, flowing through the on/off laminating valve. The reduction of pressure decreases the CO_2 density and the fluid loses some of its solvating power. Now, the CO_2 is in a subcritical state (4.93 MPa or 50 bar). Solvent vaporization is achieved by circulation of warm water in the jacket of the separator.
4. Gaseous CO_2 returns to the condenser and is liquefied again.

3.1.3. End of the Extraction

1. Stop the pump and turn off valve [V3] from the pump to the extractor.
2. Open valve [V5] slowly in the lower part of the extractor.
3. Open valves [V6] and [V7] slowly.
4. Unscrew the lids of the extractor and the separator.

5. Open valve [V8] slowly in the lower part of the separator and collect the extract (*see* **Notes 3** and **4**).
6. After each extraction, the equipment was washed for 30 minutes with *n*-hexane.

3.2. Analysis of the Extract

Several different methods can be used to analyze the extracts (*see* **Note 5**).

3.2.1. Thin-Layer Chromatography

TLC plates: Merck Silica Gel 60 (10 cm × 20 cm). Eluent system I: *n*-hexane:ethyl acetate (80:20 v/v) and 1% acetic acid. Eluent system II: *n*-BuOH:acetic acid:water (4:1:5 v/v). TLC is usually monitored at 254 and 365 nm.

3.2.2. Reverse Phase High Performance Liquid Chromatography

Analytical column chromatography [RP 8 LiChrospher 5 µm Merck (250 mm × 4 mm ID)] with gradient elution. Eluent A: 1-octane sulfonic acid sodium salt (10 m*M*) in water plus triethylamine (0.15 *M*) and acetonitrile (80:20 v/v), pH 2.5 with H_3PO_4. Eluent B: 1-octane sulfonic acid sodium salt (10 m*M*) in water plus triethylamine (0.15 *M*) and acetonitrile (40:60 v/v), pH 2.5 with H_3PO_4. Gradient program is 0 to 3 min 100% A; 3- to 28-min linear gradient at 100% B; 28- to 35-min 100% B (**Fig. 2**).

3.2.3. Gas Chromatography-Mass Spectrometry

GC:Varian 3400 equipped with injector split/splitless 250°C, split ratio 40:1, gas carrier helium, pressure gas carrier 5 psi, capillary column RSL 300 Alltech (30 m × 0.32 mm ID), film thickness 0.3 µm. Temperature program: 0 to 3 min at 80°C; 80 to 280°C with increase rate 10°C/min; isotherm 280°C for 5 min (**Fig. 3**). MS Finnigan MAT TSO-70, EI (Electron Impact), 70 eV, 200 µA, source temperature 150°C and temperature GC/MS 280°C (**Fig. 4**).

4. Notes

1. *E. californica* seeds are globular in shape, light and small (1.3 to 1.5 mm in diameter). The seeds are ground to facilitate the diffusion of the fluid into the matrix and to enhance the extraction of the analytes.
2. At lower pressure (7.89 to 10.84 MPa), supercritical CO_2 gives a yellow extract containing mainly triglycerides. In order to enhance the CO_2 solvent strength, it is necessary to increase pressure up to 24.65 MPa. The best results are obtained working at 13.80 MPa. Extraction time does not influence the quality of the extract, but pressure exerts a marked influence on red pigment purity and recovery.
3. The extracts are dark red and stable in acidic medium (pH 2), but they are light-sensitive.
4. The extract changes color (white/yellow) in alkaline medium (pH 9) coupled with white light. A saponification reaction (NaOH 1 *N* in MeOH 70%) does not affect

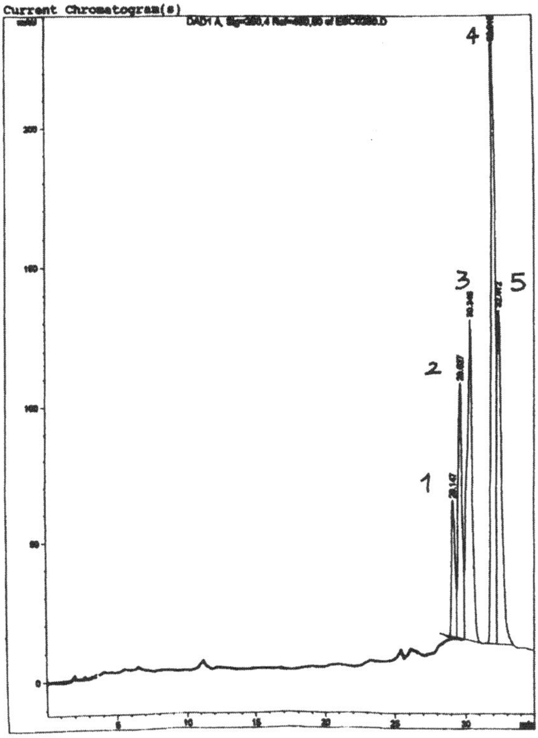

Fig. 2. HPLC chromatogram of the red pigments.

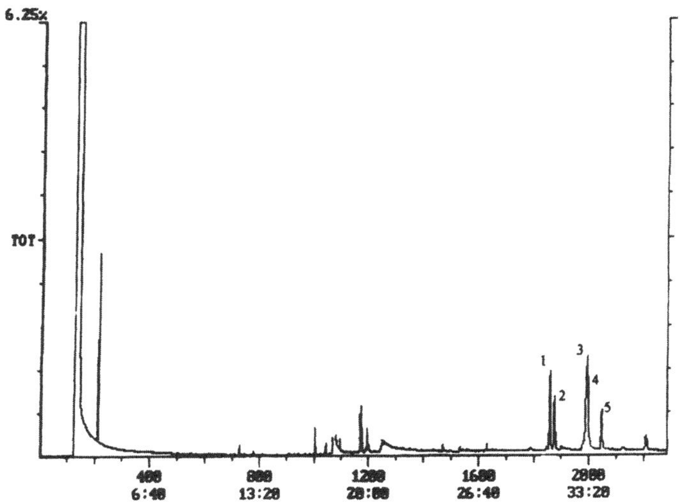

Fig. 3. GC chromatogram of the red pigments.

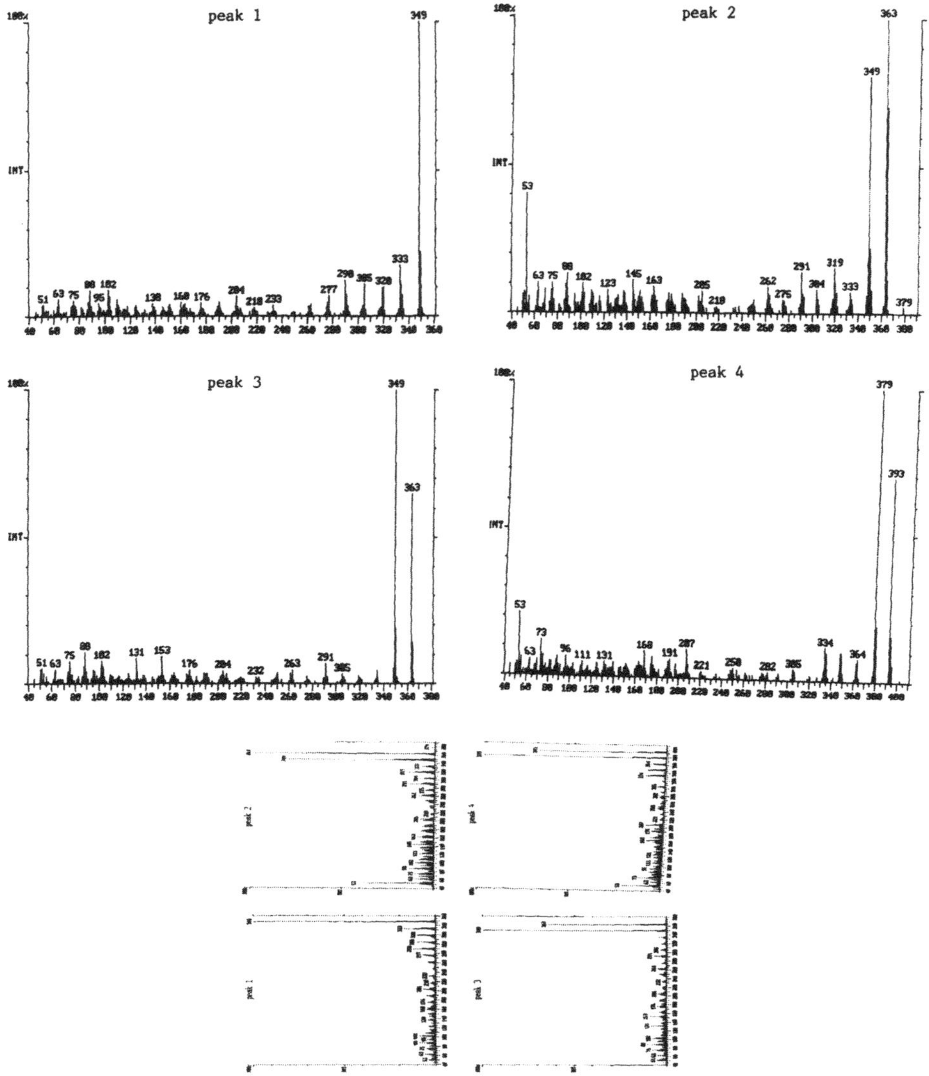

Fig. 4. MS spectra of the five isolated red pigments.

these pigments. Color reactions can play an important role in the identification of colored (red and yellow) compounds. Thus, chalcones and flavanones, for example, are isomeric and readily interconvert

5. TLC, HPLC and GC analyses all give good separations of five main red components. GC-MS analyses permitted observation of their molecular fingerprints. The red pigments could have a lactone moiety linked to an isoprene side chain, and this could be responsible for their lipophilic properties.

References

1. Duke, J. A. (1987) *Handbook of Medicinal Herbs*. CRC Press, Boca Raton, Florida.
2. Bruneton, J. (1993) *Pharmacognosie*. Tec. Doc., Paris.
3. Rolland, A., Fleurentin, J., Lanhers, M. C., Younos, C., Misslin, R., Mortier, F., and Pelt, J. M. (1991) Behavioural effects of the American traditional plant *Eschscholtzia californica*: sedative and anxiolytic properties. *Planta Med.* **57,** 212–216.
4. Dopke, W. and Fritsch, G. (1970) Alkaloid content of *Eschscholtzia californica. Pharmazie* **25,** 203–204.
5. Sarkany, S., Kovacs, A. Z., Nyomarkay, K. M., and Kerekes-Liszt, K. (1973) Fine structure and storage function of the radicle and young "seedling" root of some dicotyledonous plants. Proc. Symp. Slovak Acad. Sci., Bratislava, Czechoslovakia, 53–65.
6. Colombo, M. L. and Tomè, F. (1993) Alkaloid production during plantlets development of *Eschscholtzia californica* Cham. *Pharmacol. Res.* **27,** 5–6.
7. Bugatti, C., Colombo, M. L., and Tomè, F. (1994) Phytochemical and biological aspects of *Eschscholtzia californica* Cham. seeds. International Congress on Natural Products Research, Halifax, Canada, P 105.
8. Fox, G. A., Evans, A. S., and Keefer, C. J. (1995) Phenotipic consequences of forcing germination: a general problem of intervention in experimental design. *Am. J. Bot.* **82,** 1264–1270.
9. Colombo, M. L. and Mossa, A. (1996) Pigments rouges dans les graines de *Eschscholtzia californica* Cham. Colloque sur les Fluides Supercritiques: Applications aux Produits Naturels, Grasse, France, 127–132.

9

Supercritical Fluid Extraction of Flumetralin from Tobacco Samples

Fernando M. Lanças, Mário S. Galhiane, and Sandra R. Rissato

1. Introduction

Despite the great arsenal of analytical techniques, the success of detection, identification and quantitation of pesticide residues depends initially on the analyte extraction and/or concentration method. These methods are the most problematic step in the chemical analyses of real world samples. Not only is the majority of total analysis time spent in sample preparation, but it is also the most error-prone and the most labor-intensive task in the laboratory *(1)*. The target analyte to be separated from the matrix is usually taken up by an auxiliary substance such as a gas, a solvent and an adsorbent. These separation processes can be regarded as extraction procedures performed with liquid solvents and either a Soxhlet apparatus or sonicator. These extractions may require several hours or even days to perform, use large volumes of ultrapure solvents, and often fail to yield quantitative extraction and recovery of target analytes.

These concerns have been reflected in the development of alternative means of sample preparation for trace analysis, especially for chemically complex samples. During the last few years, supercritical fluid extraction or SFE (*see* Chapter 1) has received considerable attention as an extraction medium, primarily because of the economic and environmental consequences of organic solvent usage and disposal *(2,3)*. A substance that is above its critical temperature and pressure is defined as a supercritical fluid. The relatively high density (liquidlike) of supercritical fluids gives good solvent power, while their relatively low viscosity and high diffusivity (gaslike) values provide appreciable penetration into the matrix facilitating solute mass transfer from the matrix to

From: *Methods in Biotechnology, Vol. 13: Supercritical Fluid Methods and Protocols*
Edited by: J. R. Williams and A. A. Clifford © Humana Press Inc., Totowa, NJ

the fluid *(2)*. Recent reports have demonstrated the potential for using SFE as a replacement for more conventional liquid–solvent extraction techniques *(4,5)*.

The most common fluid for SFE applications is carbon dioxide. Due mainly to its low critical point, low toxicity and low cost, carbon dioxide has been used widely to extract among other things, natural products *(6)*, essential oils *(7)*, and pesticides *(8–10)*. When recoveries of the analyte are poor, the most common approach has been the addition of organic solvents (known as modifiers) to increase the polarity of the carbon dioxide. This either increases the solubility of the target analyte or causes interaction with active sites on the sample matrix in order to more efficiently displace the analyte *(11)*. The most common modifier used in SFE has been methanol due its high solvent polarity parameter *(12)*. However, the effect of modifier in terms of extraction power depends on the identity of the modifier, the analyte and the sample matrix. Recently, several modifiers have been investigated for the extraction of different analytes from sample matrices including pentane in the extraction of food *(13)* and PAHs *(14)*, acetone *(15)*, and *n*-hexane *(5)* in the extraction of pesticide residue. All showed improved recoveries of the target analyte when compared with pure carbon dioxide.

In addition, the system for the collection of the analytes plays an important role in obtaining efficient quantitative results in SFE. A wide variety of methods for trapping analytes have been reported, including collection in liquid solvent *(16)*, collection on adsorbent resin traps *(17)*, collection on cryogenically cooled surfaces *(18)*, and collection directly onto chromatographic columns via on-column or split/splitless injection ports *(19)*.

In the present chapter, an SFE protocol is described for the extraction of spiked flumetralin from tobacco samples. The extracts obtained were analyzed by capillary gas chromatography with electron capture detection (GC-ECD).

2. Materials

1. *n*-Hexane (pesticide grade).
2. Carbon dioxide (siphoned), SFE grade.
3. Nitrogen (ultrapure grade).
4. Solid-phase extraction cartridges, Florisil 100–120 mesh, J. T. Baker (6 cc–1 g) or equivalent.
5. Supercritical fluid extraction (SFE) system as displayed in **Fig. 1**.
6. Hydrogen used as carrier gas, ultrapure grade (99.9995%).
7. The tobacco leaves are ground, sieved in a granulator of 60 mesh and stored in a freezer (–18°C) until extraction.
8. Safety glasses and gloves should be used to work with supercritical fluids due to the high pressure used for the extraction.
9. Flumetralin (I), analytical standard, purity > 99.5% (*see* **Note 1**).

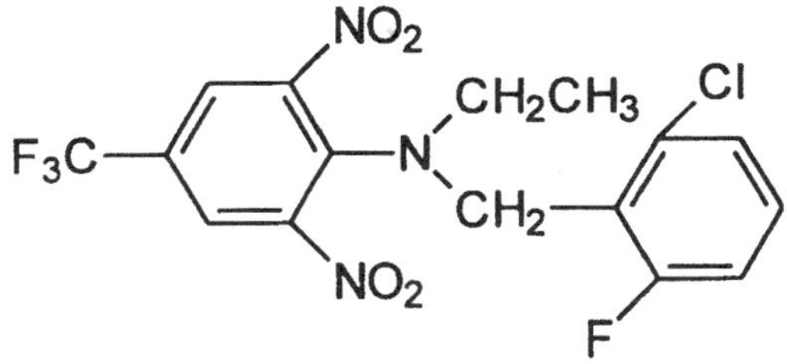

(I)

3. Method

1. The apparatus used for dynamic SFE is shown in **Fig. 1** (*see* **Note 2**). Siphoned carbon dioxide is pressurized to the required level by a Varian 8500 syringe pump (*see* **Note 3**). A 1-g sample of powdered tobacco fortified with flumetralin (*see* **step 4** of this section) is placed inside a stainless steel home-made SFE cell (*see* **Notes 4** and **5**). Before extraction, the modifier (*n*-hexane) is added to a premixture chamber by pipetting a calculated volume in relation to the total volume (64 mL) of the SFE cell so that the extraction fluid is carbon dioxide/*n*-hexane (80:20 v/v).
2. The extraction cell temperature is reached by placing both the premixture chamber and the extraction cell, connected in series by a coil transfer line of stainless steel tubing (2 m long and 1/16 in ID), inside an oven of a gas chromatograph or equivalent. To avoid a pressure buildup during the heating step, the cell is pressurized to 50 atm, with the outlet valve in position off until the extraction temperature is stable at 60°C (*see* **Note 6**). Once the extraction temperature is reached, the inlet valve is opened gently and the extraction cell pressurized to 100 atm (*see* **Note 7**). The outlet valve is opened quickly, while a linear restrictor maintains a constant pressure and controls the extraction flow rate at 160 mL/min (*see* **Note 8**). The extraction is performed for 2 min. The extract is collected in a 5 cm × 20 cm screw cap glass vial (*see* **Note 9**), specially adapted to this type of collection (*see* **Note 10**). Collection is in 20 mL of *n*-hexane at room temperature. After collection, the extract is concentrated to 1 mL and transferred to a screw cap amber vial by washing the collector with 3 × 3 mL of *n*-hexane. The extract is dried under a nitrogen flow, diluted to 3 mL of *n*-hexane and subjected to a clean-up step.
3. The clean-up step employed in this work is based on the use of solid phase extraction and is applied independently of the extraction method (*see* **Note 11**). The extract dissolved in 3 mL of *n*-hexane is applied to a Florisil cartridge (J. T. Baker) after a prewashing step with 10 mL of *n*-hexane. Flumetralin is eluted from the cartridge with 30 mL of *n*-hexane. The extract is concentrated in a rotary evaporator under reduced pressure at 50°C and submitted for analysis by GC-ECD (*see* **Note 12**).

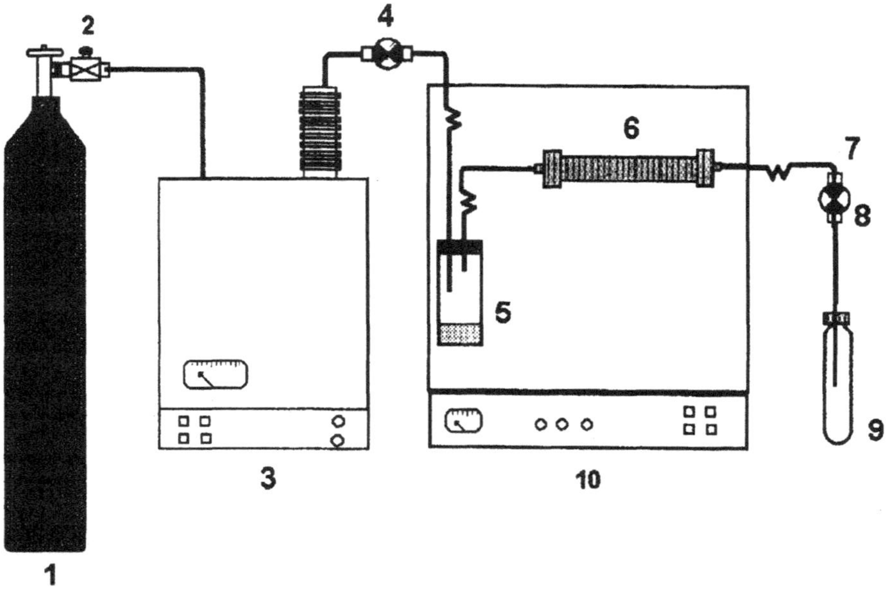

Fig. 1. Supercritical fluid extractor (dynamic mode): *1,* CO_2 tank; *2,* pressure regulator; *3,* high-pressure pump; *4,* inlet valve; *5,* mixture cell containing the modifier; *6,* extraction cell; *7,* restrictor; *8,* outlet valve; *9,* collection cell; *10,* oven.

4. Recovery of flumetralin by the SFE method described here is evaluated by the fortification of 1 g of untreated tobacco sample with 0.1 mg/L of flumetralin standard. The stock flumetralin standard solution (1 mg/L) is made by weighing 0.1 mg of standard flumetralin and solubilizing it in 100 mL of *n*-hexane (*see* **Note 13**). The other standard solutions are obtained from dilution of the stock solution with *n*-hexane. The extraction is carried out five times for calculation of the relative standard deviation of the results. A typical value obtained for flumetralin recovery by SFE from untreated tobacco sample fortified according to the standard procedure (*see* **Note 14**) is 105.3 ± 3.5%. This compares favorably to conventional solvent extraction (*see* **Note 15**) which recovers 98.4 ± 3.8%.

4. Notes

1. Flumetralin standard must be kept in the freezer (ca. –18°C). Due its toxicity, some precautions in relation to contact with skin and eyes and inhalation must be observed.
2. **Warning:** SFE solvent delivery, all tubes, cells, and connections must be checked periodically because high pressure is used during the extraction procedure. The use of convenient protective equipment is recommended.

3. The body of the supercritical fluid pressure pump should be chilled with a liquid cooling jacket to 20°C or less (due to the vapor pressure of carbon dioxide) 30 min before commencement of the extraction procedure.
4. Before filling the extraction cell, it is recommended that a piece of fused silica wool be inserted between the sample and the stainless steel frit, to avoid the partial or total clogging of the frit.
5. It is recommended that the extraction cell be only a third full to prevent back-flow of the sample, which normally blocks the carbon dioxide flow in the transfer line.
6. For the extraction temperature to reach 60°C, an equilibrium time of about 3 min is required, depending on the extraction cell wall thickness.
7. When the extraction is ready to start, a pressure equilibrium time of about 3 min is allowed (this procedure increases the extraction reproducibility).
8. Extraction flow rate depends on the restrictor size and internal diameter. For the system used in this work, the restrictor was made of a piece of fused silica capillary (50 cm × 0.05 mm i.d. × 0.12 mm o.d.) from Siemens München, Germany.
9. Sample collection is performed in deactivated vials about 7 times larger than the total volume collected, due to the high carbon dioxide flow rate, which generates turbulence at the end of the restrictor. Vial caps were specially adapted to the extract collector, by insertion of one entrance compatible with the restrictor external diameter and another 1/8-in. hole to allow the evaporated solvent and carbon dioxide to escape.
10. All glassware used has to be previously silanized using a hexamethyldisilazane/methanol 20% solution at 70°C overnight, and carefully washed with Extran solution (Merck or equivalent) to remove any coelutants due to the low level of the analytes.
11. Solid-phase extraction equipment should be cleaned before each experimental operation to avoid contamination. The extracts transferred to the SPE cartridges should be processed carefully in the following sequence: *1.* Prewashing step; *2.* Insertion of the extract; *3.* Addition of the elution solvent when 2 mL of extract remains in the top of the cartridge.
12. Quantitation (by GC-ECD) of flumetralin extracted from the fortified tobacco samples is done by an external standard method. The analytical curve is obtained over the range 0.2 to 2.0 mg/L, with 1-mL triplicate injection for each point on the curve. The recovery (R) of flumetralin from the fortified tobacco samples is calculated according to the following equation:

$$R_{(\mathrm{mg/kg})} = [(C \times V_f)/(m \times r)]\ 100,$$

where C is the analytical concentration obtained from the analytical curve, V_f is the dilution volume for analysis, m is the mass of tobacco, and r the method recovery. The gas chromatograph was equipped with split/splitless injection facilities and an electron capture detector (^{63}Ni). Injection is performed in the split mode with deactivated glass liner packed with 1 cm of 3% OV-1 over Chromossorb WAW/DMCS. Capillary column is a 5% 30 m long × 0.25 mm ID with a film thickness of 0.53 μm. The analytical conditions during all experimental procedure used are: detector temperature, 300°C; injector temperature, 250°C; initial temperature, 200°C (10 min); rate, 6°C/min; final temperature, 300°C (5 min); split ratio, 1:30; carrier gas, H_2 ($\mu^- = 38$ cm/s); makeup flow, N_2 (66 mL/min).

13. All the standard solutions are prepared from a freezed stock no more than 2 h before use and stored in amber vials at ambient temperature, normally 25°C.
14. Recovery study is made by homogeneous fortification of an untreated sample with 1 mL of standard solution in *n*-hexane. After the addition of the standard, a waiting time of 60 min before extraction is strongly recommended.
15. In the conventional solvent extraction method, 1 g of powdered tobacco is extracted with 60 mL of *n*-hexane for 20 min with constant stirring at room temperature and 60 rpm. The extract is filtered in a Buchner funnel and the solid washed with two 20 mL portions of *n*-hexane. The extract is concentrated to dryness in a rotary evaporator under reduced pressure at 40°C. The residue is dissolved in 3 mL of *n*-hexane and submitted to a clean-up step. For the extraction of spiked flumetralin from tobacco, it was found that SFE using carbon dioxide:*n*-hexane (80 : 20 v/v) in the dynamic mode gave comparable results to conventional solvent extraction. By carefully following the instructions described in this protocol, very good yields (>98%) and good repeatability (RSD ca. 5%) are obtained with a minimum detectable quantity of flumetralin of 0.005 mg/L.

References

1. Hedrick, J. L., Mulcahey, L. J., and Taylor, L. T. (1992) Fundamental review: supercritical fluid extraction. *Mikrochim. Acta* **108,** 115–132.
2. Camel, V., Tambuté, A., and Caude, M. (1993) Analytical-scale supercritical fluid extraction: a promising technique for the determination of pollutants in environmental matrices. *J. Chromatogr.* **642,** 263–281.
3. Gere, D. R., Knipe, C. R., Castelli, P., Hedrich, J., Randall, L. G., Schulenberg-Schell, H., Schuster, R., Doherty, L., Orolin, J., and Lee, H. B. (1993) Bridging the automation gap between sample preparation and analysis: an overview of SFE, GC, GC-MS and HPLC applied to environmental samples. *J. Chromatogr. Sci.* **31,** 246–258.
4. Lou, X., Janssen, H.-G., and Cramers, C. A. (1993) Quantitative aspects of directly coupled supercritical fluid extraction-capillary gas chromatography with a conventional split/splitless injector as interface. *J. High Resol. Chromatogr.* **16,** 425–428.
5. Yang, Y., Gharaibeh, A., Hawthorne S. B., and Miller D. J. (1995) Combined temperature/modifier effects on supercritical CO_2 extraction efficiencies of polycyclic aromatic hydrocarbons from environmental samples. *Anal. Chem.* **67,** 641–646.
6. Vilegas, J. H. Y., Lanças, F. M., Vilegas, W., and Pozetti, G. L. (1993) Off-line supercritical fluid extraction-high resolution gas chromatography applied to the study of *Moraceae* species. *Phytochem. Anal.* **4,** 230–234.
7. Vilegas, J. H. Y., Lanças, F. M., and Vilegas, W. (1994) Application of a home-made supercritical fluid extraction system to the study of essential oils. *Flavor Fragrance J.* **9,** 39–43.

8. Lanças, F. M., Rissato, S. R., and Mozeto, A. A. (1996) Off-line SFE-CGC-ECD analysis of 2,4-D and dicamba residues in real sugar cane, rice and corn samples. *J. High Resol. Chromatogr.* **19,** 564–568.
9. Lanças, F. M., Galhiane, M. S., and Barbirato, M. A. (1995) Extraction of norflurazon residues in cotton/seeds with supercritical CO_2. *Chromatographia* **40,** 432–434.
10. Lanças, F. M., Galhiane, M. S., Barbirato, M. A., and Rissato, S. R. (1996) Supercritical fluid extraction of chlorothalonil residues from apples. *Chromatographia* **42,** 547–550.
11. Hawthorne, S. B. and Miller, D. J. (1994) Direct comparison of Soxhlet and low- and high temperature supercritical CO_2 extraction efficiencies of organics from environmental solids. *Anal. Chem.* **66,** 4005–4012.
12. Janssen, J. G. M., Schoenmakers, P. J., and Cramers, C. A. (1989) A fundamental study of the effects of modifiers in supercritical fluid chromatography. *J. High Resol. Chromatogr.* **12,** 645–651.
13. Lanças, F. M., Queiroz, M. E. C., and Silvam, I. C. E. (1994) Seed oil extraction with supercritical carbon dioxide modified with pentane. *Chromatographia* **39,** 687–692.
14. Lanças, F. M., Martins, B. S., and Matta, M. H. R. (1990) Supercritical fluid extraction using a simple and inexpensive home-made system. *J. High Resol. Chromatogr.* **13,** 838–842.
15. Lanças, F. M., Rissato, S. R., and Galhiane, M. S. (1996) Off-line SFE-CZE analysis of carbamates residues in tobacco samples. *Chromatographia* **42,** 323–328.
16. Hawthorne, S. B. and Miller, D. J. (1987) Extraction and recovery of polycyclic aromatic hydrocarbon from environmental solids using supercritical fluids. *Anal. Chem.* **59,** 1705–1708.
17. Hedrick, J. L. and Taylor, L. T. (1990) Supercritical fluid extraction strategies of aqueous based matrices. *J. High Resol. Chromatogr.* **13,** 312–316.
18. Wright, B. W., Wright, C. W., Gale, R. W., and Smith, R. D. (1987) Analytical supercritical fluid extraction of adsorbent materials. *Anal. Chem.* **59,** 38–44.
19. Hawthorne, S. B., Miller, D. J., and Langenfeld, J. J. (1990) Quantitative analysis using directly coupled supercritical fluid extraction-capillary gas chromatography (SFE-GC) with a conventional split/splitless injection port. *J. Chromatogr. Sci.* **28,** 2–8.

10

Supercritical Fluid Extraction and High Performance Liquid Chromatography Determination of Carbendazim in Bee Larvae

José L. Bernal, Juan J. Jiménez, and María T. Martín

1. Introduction

Nowadays, several chemicals are being assayed to control the proliferation of the ascomycete *Ascosphera apis* in honey beehives. The presence and growing of these fungi in bee larvae causes their death and as a result of this a reduction in the number of bees is appreciated in the colony, and this also produces important economic losses, not only for the apiarists, but also for the surrounding farmers by means of decreasing pollination. Diverse fungicides have been assayed with the aim of penetrating into the larvae to avoid the fungi development, and among them, carbendazim [methylbenzimidazol-2-yl carbamate]

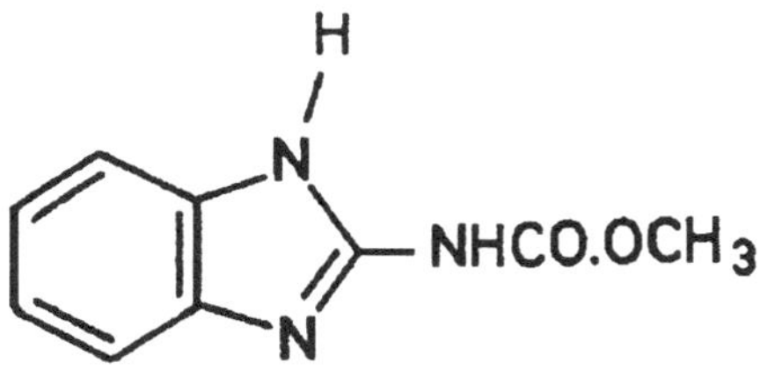

seems to be one of the most suitable chemicals to prevent the disease. To control beehive treatment and establish the therapeutic dose for the product, it is compulsory to get reliable procedures to evaluate the residues of the fungicide on larvae.

From: *Methods in Biotechnology, Vol. 13: Supercritical Fluid Methods and Protocols*
Edited by: J. R. Williams and A. A. Clifford © Humana Press Inc., Totowa, NJ

Carbendazim is a widespread fungicide and its determination on vegetables is very common. For this purpose, liquid-liquid extraction or solid–phase extraction are often used. However, applying those procedures to analyze residues of this fungicide on larvae can be difficult because of their high protein content. Consequently, those methods must be modified or alternatives sought *(1,2)*. Among the latter, supercritical fluid extraction (SFE) has been shown to be one of the most suitable *(3–6)* and the concept was introduced in Chapter 1.

To enhance the reliability and recoveries when SFE is applied on semisolid matrices, lyophilization (freeze-drying) of the sample is usually advised. In these instances, the preparation of a slurry by adding cellulose powder to the sample facilitates the lyophilization step, mainly for samples with high moisture content *(5,7,8)*. Moreover, the removal of moisture is essential to extract the carbendazim residues with supercritical CO_2. Regarding the determination of carbendazim in the extracts, reversed-phase high performance liquid chromatography (HPLC) in combination with fluorescence detection is the most frequently used technique *(1,5)*.

The purpose of this chapter is to describe a method combining SFE and HPLC with fluorescence detection to analyze carbendazim residues in bee larvae from beehives treated with the fungicide to prevent the growth of the ascospheriosis.

2. Materials

1. Lyophilization equipment furnished with a vacuum pump and a freezer system from Telstar (Barcelona, Spain).
2. A Hewlett-Packard (Avondale, PA) 7680A extractor (**Fig. 1**) equipped with a sample thimble of 7 mL and a trap packed with 550 to 650 μm stainless steel balls. Carbon dioxide (minimum purity 99.999%) is used as the extraction fluid and is supplied in cylinders with a diptube by Carburos Metálicos (Madrid, Spain).
3. The HPLC unit includes a ConstaMetric 4100 pump, an Autometric 4100 autosampler, a membrane degasser (all from LDC Analytical, Riviera Beach, FL), and a 470 fluorescence detector from Waters (Milford, MA). The column is a 150 mm × 4.6 mm Spherisorb ODS-2 from Phenomenex (Torrance, CA). The chromatography is performed under isocratic conditions with a 40:60 acetonitrile: water mixture (acidified to pH 4 with HCl) as the mobile phase. The flow rate is 1 mL/min and the temperature is 22°C. The excitation and emission wavelengths are 285 and 317 nm, respectively.
4. Carbendazim certified purity standard (99%) is purchased from Promochem (Wesel, Germany) and ultrapure water is obtained from a Milli-Q plus apparatus (Waters, Milford, MA). Hydrochloric acid is supplied by Panreac (Barcelona, Spain). HPLC-grade acetonitrile and residue analysis-grade methanol are provided by Lab-scan (Dublin, Ireland). Cellulose powder (20 μm) is obtained from Aldrich (Steinheim, Germany).

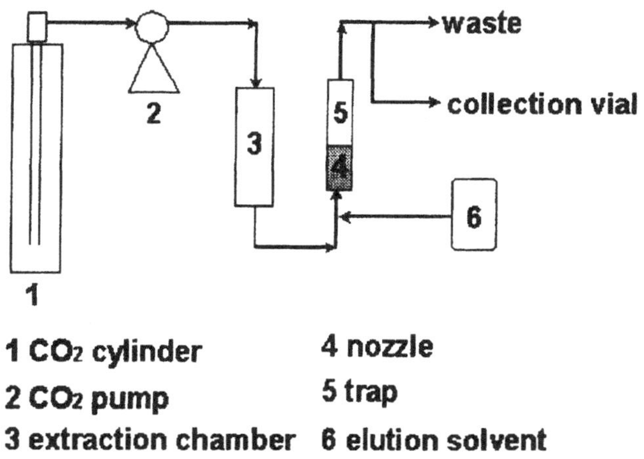

Fig. 1. Supercritical fluid extraction system: *1*, CO_2 cylinder; *2*, pump; *3*, extraction vessel; *4*, nozzle; *5*, trap; *6*, elution solvent.

5. Screw cap 14-mL vials with PTFE septa from Ohio Valley Specialty Chemical (Marietta, OH) are used, along with *autosampler 1.8-mL vials with silicone septa* from Sugelabor (Madrid, Spain). *Pipettes, glass wool, filter paper, a mortar, volumetric flasks, and other glass material* of general use are also necessary.

3. Method

3.1. Preparation of Stock Solutions

1. Weigh 50 mg of carbendazim into a 50 mL volumetric flask and fill the flask with methanol to the level (*see* **Note 1**).
2. Make a 1:10 dilution with methanol to obtain the work solution (*see* **Note 1**).
3. Dilute the work solution with methanol to obtain the HPLC calibration solutions (the standards) in the 0.50 to 15 mg/L range (*see* **Note 1**).

3.2. Extraction

1. Rinse the larvae samples with water to remove honey and beeswax residues (*see* **Note 2**).
2. Distribute about 20 g of larvae on Petri plates and freeze them at –35°C (*see* **Notes 3** and **4**).
3. Place the Petri plates in the lyophilization equipment. Pump down until constant weight is achieved, which requires about 18 h, depending on the sample size (*see* **Note 5**).
4. Mix and grind the lyophilized samples in a glass mortar.
5. Tighten an end-cap on to an extraction thimble. Place a small filter paper disk (of slightly higher diameter than that of the thimble) in the bottom of the thimble (*see* **Note 6**).

6. Weigh 0.25 g of powdered sample and place it in the bottom of the extraction thimble. Place another paper disk on the sample, put it into the thimble, and press down the sample with the help of a rod. Fill the thimble with cotton or glass wool to reduce the dead volume of the thimble (*see* **Note 6**).
7. Add 100 µL of methanol to the top of the sample in the thimble and close it (*see* **Note 7**). Place the thimble in the supercritical fluid extractor in an upside-down position.
8. Close a 1.8 mL vial with a silicone-based cap and place it on the tray rack.
9. Set the extraction conditions: fluid density 0.75 g/mL, extraction chamber temperature 50°C, operating pressure 176 bar, supercritical CO_2 flow-rate 1.5 mL/min, equilibrium time 2 min, dynamic extraction time 30 min, and nozzle temperature 75°C. Carbendazim was collected on the trap at 5°C.
10. Run the extraction.
11. After extraction, elute the trap with 1.5 mL of methanol at 45°C and a flow-rate of 0.2 mL/min (*see* **Note 8**). Pick up the vial for HPLC analysis.
12. To clean the trap, rinse it with 3 mL of methanol at 45°C and a flow-rate of 1 mL/min.

3.3. HPLC Analysis

1. Inject (20 µl) each of the carbendazim standards into the HPLC system separately.
2. Verify the linearity of detector response over the concentration range 0.50 to 15 mg/L. The correlation coefficient must be a minimum of 0.99.
3. Check the limits of detection and quantitation (LOD and LOQ) established by the equations:

$$\text{LOD} = 3 \times s_{x/y}/b,$$

and

$$\text{LOQ} = 10 \times s_{x/y}/b,$$

where $s_{x/y}$ is the standard deviation of the linear fitting and b is the slope of the fitting. An LOD and an LOQ of 0.1 and 0.25 mg/L, respectively, are easily obtained (*see* **Note 9**).
4. Make blanks: apply the SFE-HPLC method to nontreated samples to test for the presence of coextracted substances (from either the larvae matrix or the reagents) that could interfere with the chromatographic determination.
5. Inject 20 µL of extract into the HPLC system and run the chromatogram (*see* **Notes 10** and **11**).
6. Integrate the chromatogram and report the peak area of carbendazim.
7. Check the recovery of carbendazim, the (intraday) repeatability and the (interday) reproducibility of spiked samples (*see* Note **12**). Recoveries must be higher than 85% for bee-larvae samples spiked with 10–100 mg/kg. Repeatabilities and reproducibilities, as measured by relative standard deviation, must be close to 3.8 and 5.5%, respectively (seven determinations).

4. Notes

1. The carbendazim stock solutions kept in the freezer at –20°C can be used for at least 3 mo. The work solutions kept in the refrigerator at 4°C must only be used

for 1 mo. Standards for HPLC calibration must preferably be prepared every day, if no preservation precautions are taken. Stock and work solutions are kept in 14-mL vials, while calibrations standards are placed in 1.8-mL vials.

2. Always rinse the bee larvae because of the beeswax and honey residues that cover them. This step is important to avoid a loss of efficacy in the extraction and to avoid the presence of interfering peaks (arising from coextracted substances) on the chromatograms.
3. Keep the larvae samples in the freezer until the extraction, if it is not possible to extract them immediately.
4. A sample size of 20 g of fresh bee larvae is an adequate quantity to achieve a representative sample. However, smaller samples are often handled due to their low availability.
5. The lyophilization system freezes the sample contained in the Petri plates and pumps down at the same time. However, and as a precaution, it is convenient to introduce the sample already frozen in the lyophilization system to prevent splashing of the sample when the vacuum pump is switched on.
6. The disks of filter paper help to keep the sample powder together in addition to preventing the powder reaching the thimble caps, which could cause blockages. Furthermore, the insertion of glass or cotton wool between the paper disks and caps is sometimes advisable to reduce blockages. The glass or cotton wool must be rinsed with methanol before use. Maintaining the screwed portion of the thimble free of particles helps to prevent leaks.
7. The addition of methanol as modifier is necessary to obtain high carbendazim recoveries and acceptable precisions, mainly for the extraction of samples containing high carbendazim concentrations.
8. To obtain reproducible results, the stainless steel ball trap must be rinsed with abundant methanol, at least 20 mL, before its daily use.
9. Detection limits of about 0.1 to 0.25 mg/kg are usually obtained. Quantitation limits are close to 0.3 to 0.6 mg/kg. These limits are sufficient for the analysis of carbendazim residues found on larvae from treated beehives. It should be taken into account that just after dosing with carbendazim, larvae can have concentrations of the fungicide of up to 100 mg/kg (fresh weight).
10. The elution of carbendazim through the HPLC system equipped with an ODS column is very sensitive to the presence of active sites when the pH of the mobile phase is close to 7. Also, in this case, the chromatographic system was roughly stabilized; a higher retention time for carbendazim was observed from run to run. The acidification of the mobile phase up to pH 4 completely solves those problems. The system equilibrates in a few minutes and the retention time and area for carbendazim are reproducible. When the pH of the mobile phase is lowered, longer retention times are observed. Retention can be shortened by increasing the percentage of organic modifier in the mobile phase. In the aforementioned operating conditions, a retention time of 4.0 min is achieved for carbendazim.
11. Dilute the extract with methanol to reach the calibration range if high concentrations of carbendazim are expected.

12. The use of carbendazim-spiked samples is necessary to carry out reliable studies about the recovery achieved by the procedure due to the lack of certified reference samples. Studies of precision can be made either on spiked or real samples. Fortification is made before the lyophilization step. For this purpose, an amount of larvae, for instance 20 g, is ground in a glass mortar and spiked with 1 mL of an aqueous solution containing carbendazim of known concentration. The spiking of the sample just before the start of lyophilization is favored by grinding the sample, which is necessary to allow the fungicide to soak into the matrix. To reduce splashing, about 2 to 3 g of cellulose powder are added to the slurry. The slurry is vigorously homogenized by manual shaking and frozen before lyophilization.

Acknowledgment

Larvae samples were kindly supplied by Mr. Mariano Higes from Centro Apícola Regional of Marchamalo (Guadalajara, Spain).

References

1. Bernal, J. L., del Nozal, M. J., Toribio, L., Jiménez, J. J., and Atienza, J. (1997) High performance liquid chromatography determination of benomyl and carbendazim residues in apiarian samples. *J. Chromatogr. A* **787,** 129–136.
2. Bernal, J. L., del Nozal, M. J., Rivera, J. M., Jiménez, J. J., and Atienza, J. (1996) Determination of the fungicide vinclozolin in honey and bee larvae by solid-phase extraction with gas chromatography and electron capture and mass spectrometric detection. *J. Chromatogr. A* **754,** 507–513.
3. Lee, M. L. and Markides, K. (1990) *Analytical supercritical fluid chromatography and extraction*. Chromatography Conferences, Provo, Utah.
4. Majors, R. E. (1992) Fundamental considerations for SFE method development. *LC-GC Int.* **5,** 8–10.
5. Jiménez, J. J., Atienza, J., Bernal, J. L., and Toribio, L. (1994) Determination of carbendazim in lettuce samples by SFE-HPLC. *Chromatographia* **38,** 395–399.
6. Aharonson, N., Lehotay, S. J., and Ibrahim, M. A. (1994) Supercritical fluid extraction and HPLC analysis of benzimidazole fungicides in potato, apple and banana. *J. Agric. Food. Chem.* **42,** 2817–2823.
7. Atienza, J., Jiménez, J. J., Bernal, J. L., and Martín, M. T. (1993) Supercritical fluid extraction of fluvalinate residues in honey: determination by high performance liquid chromatography. *J. Chromatogr. A* **655,** 95–99.
8. Atienza, J., Jiménez, J. J., Alvarez, J., Martín, M. T., and Toribio, L. (1994) Extraction with EDTA/methanol and supercritical carbon dioxide for the analysis of ziram residues on spinach. *Toxicol. Environ. Chem.* **45,** 179–187.

11

Supercritical Fluid Extraction Coupled with Enzyme Immunoassay Analysis of Soil Herbicides

G. Kim Stearman

1. Introduction

The purpose of this chapter is to describe a supercritical fluid extraction (SFE) method coupled with enzyme immunoassay analysis (EIA) for the determination of the herbicides: 2,4-dichlorophenoxyacetic acid (2,4-D), simazine, atrazine, alachlor, and trifluralin in soil.

1.1. SFE Theory and Procedure

The SFE of organics from various environmental matrices has been utilized recently to avoid using large amounts of hazardous organic solvents, commonly used in traditional extractions. The basic principles of SFE were introduced in Chapter 1. SFE, when coupled with EIA of the extracted pesticides, requires negligible organic solvent consumption and offers an alternative, inexpensive, safe and environmentally compatible method for determining pesticides in soil samples. The major problems with SFE are as follows:

1. Extraction cells must be uniformly packed.
2. The system may leak.
3. Restrictor flow may not be uniform.
4. Restrictors may clog.

Once the SFE method parameters are experimented with and it is determined what is successful on a particular system, the above-mentioned problems can largely be prevented.

The SFE of pesticides from soil often requires the addition of polar organic modifiers, such as acetone or methanol, to supercritical carbon dioxide (CO_2). The purpose of the modifier can be twofold; to increase the solubility of the

From: *Methods in Biotechnology, Vol. 13: Supercritical Fluid Methods and Protocols*
Edited by: J. R. Williams and A. A. Clifford

analyte and/or to increase the surface area of the soil, by swelling the matrix (soil) or to competitively adsorb with the analyte to the soil. Extraction temperature must be increased as the modifier percentage is increased, in order to maintain the mixture in the supercritical state. Modifiers can also be added directly to the soil in the extraction cell as entrainers *(1)*.

The concentration of the pesticides in the soil can be important in determining extraction recoveries, as there may be differences in recovery between pesticide concentrations of 10 ppm vs 10 ppb under identical conditions. This may be due to the fact that at lower analyte concentrations, a larger percentage of the total pesticide concentration is less accessible to the extraction solvent than at higher pesticide concentrations.

In addition to the actual extraction of the analyte from the matrix, the mode of sample collection plays an important role. Collection can be achieved either by directly eluting the sample into a liquid or by trapping on a solid phase, followed by solvent desorption.

A method to quantify pesticides in soil, which combines SFE and EIA, is explained. This technique limits the amount of solvents used and reduces the time of analysis compared to traditional gas or liquid chromatography.

1.2. EIA Theory and Procedures

EIA theory is based on antibody coating of microwells that allow only certain distinct compounds to bind with them, so that it is extremely sensitive and specific to the analyte of interest. The analyte competes with the enzyme conjugate for the limited number of binding sites on the microwells. The amount of enzyme conjugate that binds with the antibody is measured colorimetrically and is inversely proportional to the concentration of the analyte *(2)*. EIA has gained acceptance as a technique for the rapid determination of pesticides *(3)*. EIA can be used both as a screening method and as a semiquantitative method depending on the history of the sample. EIA microtiter plate techniques are simple to use and 40 samples can be analyzed simultaneously. In many cases, EIA is also less expensive than traditional GC or HPLC methods. The major problem with using the EIA technique is the cross reactivity of similar compounds, i.e., triazine compounds will cross-react with varying degrees of sensitivity with the EIA designed for atrazine *(4)*. This is not a problem with soil that contains no cross-reacting compounds and that is spiked and extracted shortly after spiking. However, with field-weathered samples, the metabolites can, in some cases, be more sensitive to EIA than the parent compound *(5)*.

2. Materials

1. The SFE unit includes an oven, a pump to achieve high pressure, extraction cells, and collection vials with appropriate tubing. The apparatus was an SFE Model

703 with modifier pump (Dionex, Sunnyvale, CA). This unit is designed to extract 8 cells simultaneously.
2. SFE cells are available in sizes from 0.5 mL to 10 mL.
3. SFE collection vials containing a glass tube with a C_{18} solid-phase disk, which is inserted in the septum of the collection vial cap, are used for trapping pesticides (*see* **Note 1**).
4. Hydromatrix (Varian, Inc., CA) is recommended for use as a drying agent for wet soils.
5. EIA materials include 96 antibody-coated microwell plates, specific for atrazine, triazines, 2,4-D, alachlor or trifluralin. Included in these EIA kits are solutions of enzyme conjugate, chromogen, substrate, and stop solution of 2.5 N H_2SO_4 (kits formerly supplied by Millipore or Agri Diagnostics and now supplied by Strategic Diagnostics, Newark, DE). Stability of these kits is generally 6–12 mo at 4°C in the refrigerator.
6. Herbicide standards are made from pure pesticide analytes obtained from the manufacturer.
7. Other equipment and supplies include a microtiter plate reader or colorimeter, pipettes, orbital shaker, glass wool, and modifier solvents.
8. Purity of CO_2 gas depends on analyte and interferences. SFC-grade CO_2 with a 2000 psig helium head was used in this study (Scott Specialty Gases, Plumsteadville, PA).
9. Modifiers (HPLC grade) used included acetone, triethylamine, and reagent-grade water (*see* **Notes 2** and **3**).

3. Method

3.1. Supercritical Fluid Extraction

1. Tighten the end onto the extraction cell with a wrench.
2. Place about 0.25 to 0.5 inch of glass wool into the end of the extraction cell.
3. Add a known amount of soil (3–10 g), depending on extraction cell size. If the soil is wet, mix one part Hydromatrix with 2–4 parts soil, depending on the moisture content of the soil, before loading the cell with soil. After the addition of the soil, lightly tap the loaded extraction cell on the laboratory bench a couple of times. For this study, the soil is air-dried and ground to 2 mm. Both field-weathered and laboratory-fortified soils can be used.
4. Place the glass wool into the top end of the extraction cell.
5. Tighten the extraction cell top with a wrench so no leaks result; depends on experience with leaking cells (*see* **Note 4**).
6. Place the extraction cell(s) into the SFE oven and hand tighten to connect to the gas line and collection vial.
7. Place the collection vials into the collection tray rack with C_{18} tubes attached to septa collection vial caps (*see* **Note 1**).
8. Set parameters: oven temperature: 66°C; pressure: 3 min at 200 atm followed by 17 min at 340 atm; time: 20 min; restrictor temperature: 150°C; collection vial temperature: 4°C, and modifier (90:10:2.5, acetone, water, triethylamine, v:v:v) was added to CO_2 at 10 mole%.

9. Start the SFE program and extract the samples for 20 min using the above stated parameters (*see* **Note 5**).
10. Remove the collection vials from the tray.
11. Desorb or elute any remaining analyte on the C_{18} cartridge tubes with 2 mL of acetone into the collection vial.
12. Measure the solvent volume in the collection vial.

3.2. Enzyme Immunoassay Analysis

1. Allow an EIA plate to warm to room temperature by removing it from the refrigerator at least 2 h before use. Temperature and time must be controlled for EIA to work properly. In all cases, enzyme immunoassay plates and solutions are allowed to warm to room temperature before use, and reaction times are consistent throughout the experiments (*see* **Note 6**).
2. Dilute the solvent 25-1 to 200-1 depending on expected concentration by taking an aliquot and diluting with reagent-grade water. The EIA will not tolerate acetone above 5%.
3. Make the standards in acetone at the same dilution as the unknowns (the range of standards is dependent on the analyte and is specified by the EIA kit).
4. Add 80 µL of standard or unknown to each of two microwells and proceed with samples. A partial plate or full 96-well plate may be used (*see* **Note 7**).
5. Add 80 µL enzyme conjugate to each microwell.
6. Cover plate with paraffin film to prevent spillage.
7. Mix on an orbital shaker at 200 rpm for 60 min, depending on kit (follow specific kit instructions).
8. Take off the shaker and pour the contents out and rinse the microtiter wells with deionized water five times.
9. Add chromogen and substrate using a multichannel pipette (8 rows simultaneously) to wells and let the blue color develop by mixing on an orbital shaker for 30 min. The darker the blue color, the less the analyte.
10. Take the samples off the shaker and add the stop solution using the multichannel pipette. This turns the blue solution yellow and stops the reaction.
11. Mix on the orbital shaker at 200 rpm for 5 min.
12. Read on a microtiter plate reader at 450 nm or specified wavelength. The color is stable for about 1 h (*see* **Note 8**).
13. Compute a standard curve and use it to determine the unknowns (*see* **Note 9**).
14. Compute the recoveries of the pesticides (*see* **Note 10**).

4. Notes

1. Liquid collection of analyte results in loss of analyte through aerosoling or volatilization. To prevent this loss, solid-phase C_{18} disks are used to trap analyte.
2. Restrictor clogging occurs especially with methanol as modifier (with other SFE units this may not be a problem). This is corrected by using acetone in place of methanol. Also, by ramping pressure up to the desired level, restrictor clogging is reduced.

3. It was necessary to add modifier to the CO_2, because CO_2 by itself did not achieve sufficient recovery of these analytes.
4. Possible leaks from extraction cells or connective hook-ups. This is solved by making sure that fittings and extraction cells are properly secured, largely through the experience of the operator.
5. Each series of eight extractions requires about 1 h to load, extract and elute the samples.
6. The EIA procedure requires about 2½ h for one full plate (40 samples and 8 standards in duplicate).
7. When adding sample to plates, keep track of what wells have been filled. It may be easy to get confused as to whether a well has had solution added to it or if it is the correct sample number, if a system is not set up to determine this. Set up a system so that you can keep track of your progress and make sure you can backtrack to determine the steps taken.
8. EIA color stability is about 1 h. We generally analyze samples immediately upon removal from shaker.
9. The SFE-EIA method has detection limits of 2.5 ng/g soil for atrazine and alachlor, and 15 ng/g soil for simazine and 2,4-D, without concentration of sample.
10. With this method, we have achieved recoveries of above 80% with less than 15% relative standard deviation (RSD) for 2,4-D, simazine, atrazine, and alachlor. Atrazine and alachlor recoveries have been above 90% with RSD below 10%. Atrazine and alachlor are more sensitive (lower standards used) to their respective immunoassay kit than 2,4-D and simazine. Trifluralin is not successfully analyzed by EIA because cross-reacting metabolites are more sensitive to the antibody than trifluralin. Trifluralin is analyzed by gas and liquid chromatography. The same SFE conditions are used to extract more than 85% of trifluralin from spiked and field samples. Using this SFE-EIA method, it is possible to extract and analyze 40 soil samples in an 8-h day.

References

1. Stearman, G. K., Wells, M. J. M., Adkisson, S. M., and Ridgill, T. E. (1995) Supercritical fluid extraction coupled with enzyme immunoassay analysis of soil herbicides. *Analyst* **120,** 2617–2621.
2. Stearman, G. K. and Adams, V. D. (1992) Atrazine soil extraction techniques for enzyme immunoassay microtiter plate analysis. *Bull. Environ. Contam. Toxicol.* **48,** 144–151.
3. Kaufman, B. M. and Clower, M. Jr. (1995) Immunoassay of pesticides: an update. *J. AOAC Intl.* **78,** 1079–1090.
4. Thurman, E. M., Meyer, M., Pommes, M., Perry, C. A., and Schwab, A. P. (1990) Enzyme linked immunosorbent assay compared with gas chromatography/mass spectrometry for the determination of triazine herbicides in water. *Anal. Chem.* **62,** 2043–2048.
5. Stearman, G. K. and Wells, M. J. M. (1993) Enzyme immunoassay microtiter plate response to atrazine and metolachlor in potentially interfering matrices. *Bull. Environ. Contam. Toxicol.* **51,** 588–593.

12

The Supercritical Fluid Extraction of Drugs of Abuse from Human Hair

Pascal Kintz and Christian Staub

1. Introduction

1.1. Hair as a Biological Specimen

The presence in the body of drugs of abuse can be identified by a variety of laboratory procedures. The standard in drug testing is the immunoassay screen, followed by a gas chromatography/mass spectrometry (GC/MS) confirmation conducted on a urine sample. In general, drug concentrations in urine can be determined only when exposure to the drugs occurs 2–4 d before sample collection. In recent years, remarkable advances in sensitive analytical techniques have enabled the analysis of drugs in unconventional biological samples such as hair. Since 1979, hair has been used to document chronic human drug exposure *(1)*. To date, more than 400 articles concerning hair analysis have been published *(2)*, with applications in clinical *(3)* and forensic *(4)* toxicology.

Hair is a product of differentiated organs in the skin of mammals. It differs in individuals only in color, quantity, and texture. Hair seems to be a vestigial structure in humans, since it is too sparse to provide protection against cold or trauma. Hair composition is primarily protein, but also water and lipids. The total number of hair follicles in an adult man is estimated to be about 5 million, with 1 million found on the head. Each hair shaft consists of an outer article that surrounds a cortex. The cortex may contain a central medulla.

Hair shafts originate from follicles that have various periods of activity. A follicle that is actively producing hair is said to be in the anagen phase. After a period of activity during which hair is continuously produced, at a rate in the range 0.22 to 0.52 mm/d, the follicle enters in a relatively short transition period of about 10 wk, known as the catagen phase, during which it begins to shut

From: *Methods in Biotechnology, Vol. 13: Supercritical Fluid Methods and Protocols*
Edited by: J. R. Williams and A. A. Clifford © Humana Press Inc., Totowa, NJ

Table 1
Comparison Between Urine and Hair for Testing Drugs

Parameters	Urine	Hair
Drug	Metabolites > parent drug	Parent drug > metabolites
Detection period	2–4 d	Months to years
Storage	–20°C	Ambient temperature
Screening technology	Immunoassay	GC/MS
Confirmation technology	GC/MS	GC/MS
Analysis duration	+	+++
Cost per unit test	+	+++
Adulteration	Possible	Unknown
Degree of drug use	No	Yes
Pattern of drug use	No	Yes
Quality control	Yes	Yes

down in preparation for an inactive or quiescent period, known as the telogen phase. On the scalp of an adult, approximately 85% of the hair is in the growing phase and the remaining 15% is in a resting stage *(5,6)*.

The mechanism by which drugs are deposited into hair is not well understood. Both incorporation during hair growth from the bloodstream and incorporation after hair growth from sweat and external contamination have been proposed to account for drugs appearing in hair *(7)*.

The major practical advantage of hair testing compared with urine testing for drugs is its larger surveillance window: weeks to months, even years, depending on the length of the hair shaft, versus a few days. In fact, for practical purposes, the two tests complement each other. Urine analysis provides short-term information of an individual's drug use, whereas long-term histories are accessible through hair analysis *(8,9)*.

A comparison between urine and hair is presented in **Table 1**. One of the main advantages of hair is that multisectional analysis can be performed, which consists of taking a length of hair and cutting it into sections to measure drug use during shorter periods. This technique can be applied to provide a retrospective calendar of an individual's drug use.

1.2. Analytical Tools for Drug Testing in Hair

Collection procedures for hair analysis for drugs have not been standardized. Hair is best collected from the area at the back of the head, called the vertex posterior. Compared with other areas of the head, this area has less variability in hair growth rate, the number of hairs in the growing phase is more constant and the hair is less subject to age and sex-related influences *(10)*. The sample size varies considerably among laboratories, ranging from a single hair to 200 mg, but samples of 30–50 mg are currently used ***(11)***.

Storage can be achieved at ambient temperature until analysis. The analytical procedures generally involve the following steps:

1. Decontamination of the specimen
2. Preparation (such as pulverization or segmentation in 2–3 mm)
3. Hydrolysis (acid or alkaline or enzymatic …)
4. Purification (extraction, concentration, and derivatization …)
5. Analysis by chromatography

After the third step, immunoassay screening is possible. **Step 3** and **step 4** can be combined when using methanolic incubation. GC/MS represents the state-of-the-art for the identification and quantification of drugs in hair, owing to its separation ability, detection sensitivity, and specificity. Several analytical reviews have been published to document the analytical procedures that were reported in the literature ***(12–15)***.

1.3. Importance of Supercritical Fluid Extraction in Hair Analysis

Of the articles that addressed analyses of hair, almost all present three separate stages before GC/MS, including a washing stage, a pretreatment stage, and an extraction stage. In 1995, Cirimele et al. ***(16)*** proposed a unique procedure based on supercritical fluid extraction (SFE) that allows direct preparation of the specimen for GC/MS in less than 1 h. However, it was Sachs et al. in 1992 ***(17)*** and in 1993 ***(18)*** that demonstrated for the first time the use of supercritical fluids for the extraction of drugs from hair. They illustrated the possibility of extracting opiates and cocaine from hair by means of a mixture of CO_2-ethyl acetate. Heroin, the parent drug, was identified for the first time, and therefore SFE was presented as a soft analytical tool avoiding decontamination, pretreatment, and extraction. However, recovery of the extraction remained inferior to other conventional techniques. Major improvements were obtained 2 yr later by Edder et al. ***(19)*** who demonstrated the quantitative extraction of opiates from hair. They established the composition of the polar modifier, that is, methanol:triethylamine:water (2:2:1, v/v/v) that was also used by Cirimele et al. ***(16)*** and Morrison et al. ***(20)***, but in some different proportions. More recently, Staub et al. ***(21)*** in their review documented successful applications of SFE to the screening of opiates, cocaine and methadone in human hair obtained from drug addicts.

All the authors involved in the SFE of drugs from hair have made enthusiastic comments: SFE avoids the use of environmentally damaging substances, SFE can be automated and coupled on-line with chromatographic systems such as GC/MS, SFE is faster than other traditional methods of sample preparation and SFE can be used as a screening procedure.

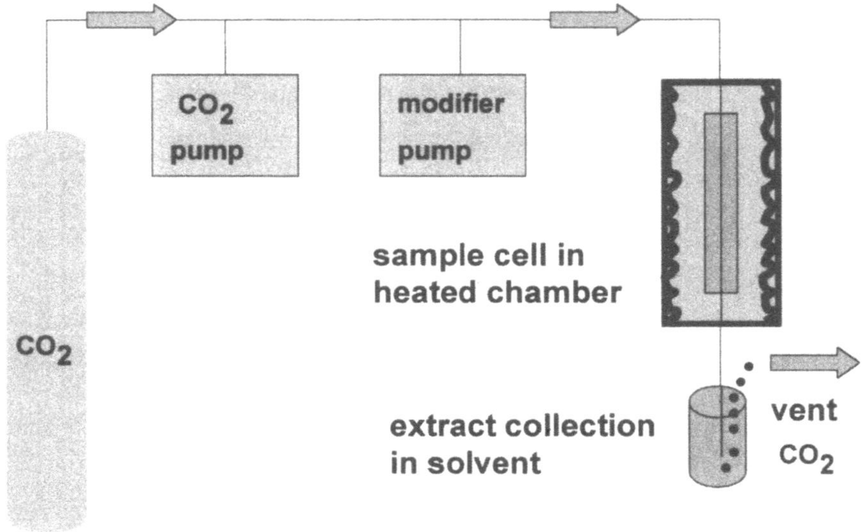

Fig. 1. Labeled diagram of the extractor.

2. Materials

1. The hardware required is relatively simple (*see* **Fig. 1**): a source of high purity supercritical fluid, a high-pressure pump capable of delivering fluid at a constant controllable pressure, an extraction chamber with a suitable heating mechanism, a restrictor to maintain pressure within the extraction chamber, and a container to collect the extracted drugs.
2. Source of supercritical fluid: CO_2 suitable for SFE is available in various tank sizes, purities and tank types. From our experiment, we recommend the use of tanks with helium headspace and a diptube, because pump filling is fast and nearly complete even without additional cooling.
3. The addition of the modifier can be made in different ways. If simple impregnation of the hair before SFE is used, important concentration gradients could occur if working in dynamic mode. It is also possible to directly use a cylinder containing the mixture, but in spite of the elegance of the method it has been shown that the mixture of polar modifier becomes richer during the emptying of the cylinder. It is therefore recommended that a second pump be used along with a mixing chamber in order to obtain a homogeneous mixture. The organic modifier should be of HPLC or GC purity.
4. The extraction vessels: because of the necessity of working at high pressures, the resistance to pressure and the absence of leaks are the principal characteristics that an extraction vessel should have. While most SFE instrumentation requires the use of vessels resistant to high pressures, Isco (Lincoln, NE) has found a way around this problem. The vessel is placed in a chamber under high pressure, and

thus both the inside and the outside of the vessels are under pressure simultaneously. It is no longer necessary, therefore, to use expensive stainless steel vessels.

5. Flow control device: as mentioned previously, the SFE pump should produce a constant pressure of supercritical fluid with a rate controlled by a flow restrictor after the extraction vessel. A self-heated needle valve restrictor is recommended (*see* **Note 1**). This kind of restrictor provides adjustable flow and accommodates liquid trapping without carry-over and plugging limitations. Heating is concentrated at the orifice and in the valve body assembly. Flow rates from 0.5 to 10 mL/min are available by just turning the control knob. Temperatures up to 150°C may be programmed into the restrictor controller.
6. The analyte collection is made by liquid trapping of extracts into a 10-mL glass tube, containing methanol as a collection solvent.

3. Methods

3.1. Specific Analysis of Opiates in Human Hair

This method can be used for quantitative analysis of the following three opiates: morphine, codeine, and 6-monoacetylmorphine (*see* **Fig. 2** for the structures of the compounds).

1. Before SFE, the hair is briefly washed by percolation with 10 mL of methylene chloride, 10 mL of water, and, finally, 10 mL of methanol.
2. After this decontamination, the drug abuser's hair is pulverized for 10 min with a ball-mill purchased from Retsch (Schieritz, Hauenstein, Switzerland).
3. Standard soaked hair is prepared by adding an aqueous standard drugs solution to the pulverized hair. The mixture is submitted to magnetic stirring for 5 h, then filtered and the hair is washed with water and methanol.
4. The extraction cell is filled with pulverized hair (about 50 mg) and placed in the extraction chamber.
5. The oven is heated at 40°C and the restrictor is heated at 80°C. At this temperature, we are using a subcritical fluid.
6. The modifier pump is filled with the following modifier mixture: methanol–triethylamine–water (2:2:1 v/v/v).
7. The hair samples are then extracted with 15% of the modifier in CO_2. The SFE conditions are the following: pressure: 250 atm, flow-rate: 1 mL/min, extraction time: 30 min or a 30-mL volume of extraction fluid.
8. During the extraction, the opiates are trapped in 10 mL of methanol.
9. After SFE, the methanol is evaporated to dryness under a gentle stream of nitrogen.
10. The extracts obtained by SFE are evaporated until dried under nitrogen flow. The opiates are then derived by propionylation. After evaporation of the solvent, 100 µL of propionic anhydride (99%, Aldrich) and 100 µL of pyridine (99.5%, Merck) are added to the residue obtained and heated at 60°C for 30 min. After evaporation of the derivatization reagents under nitrogen, the residue is dissolved in 50 µL of ethyl acetate. The nalorphine, added after the SFE, is used as a chromatographic standard (*see* **Note 2**).

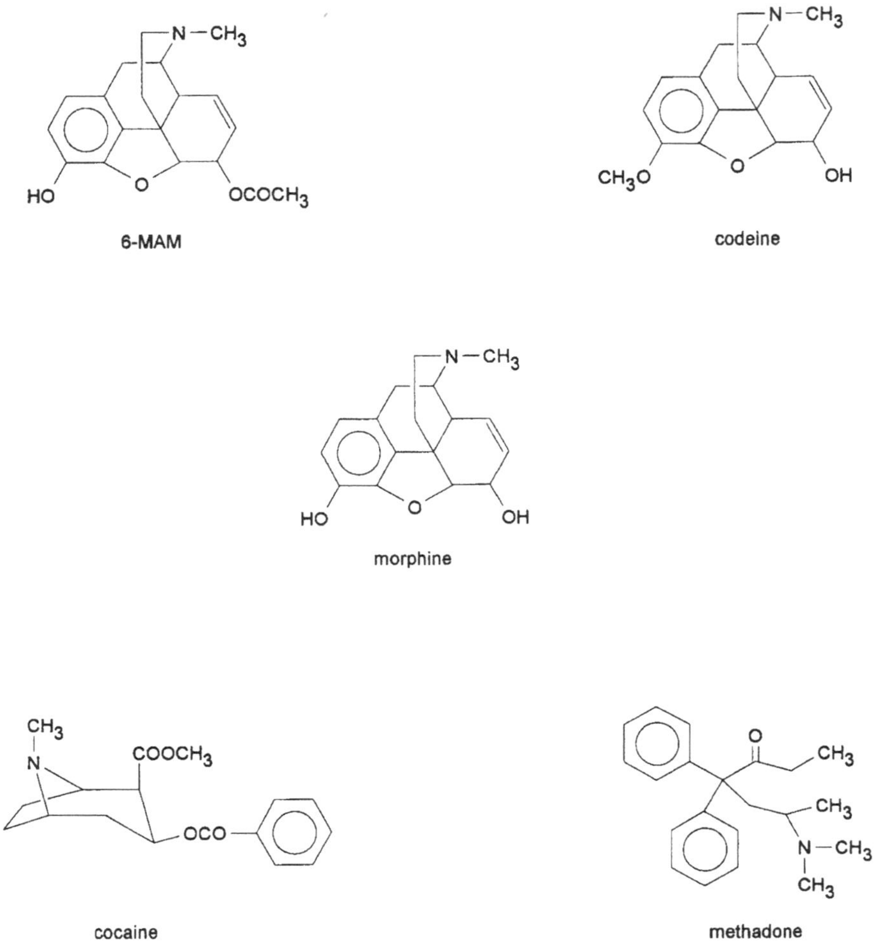

Fig. 2. Structures of the drugs.

11. The characterization and quantification of opiates are obtained with the aid of a GC/MS (*see* **Note 3**). The apparatus and the conditions used are the following:
    GC/MS: HP 5988
    Injection: splitless 3 µL at T = 270°C
    Column: DB-5 ms 15 m × 0.25 mm ID with a film thickness of 0.25 mm (J & W Scientific, Folsom, CA)
    Temperature program:
    85°C, 1 min
    190°C (15°C/min, 0.5 min)
    210°C (2°C/min, 1 min)
    270°C (20°C/min, 8 min)

Interface: 280°C
Source: 200°C
Quantitative analyses are obtained by Single Ion Monitoring (SIM):
codeine: m/z = 355 and 282, 6-MAM: m/z = 383 and 327, morphine: m/z = 397 and 341, nalorphine: m/z = 423 and 367.

3.2. Screening of Opiates, Cocaine, and Methadone in Human Hair

This method can be used for qualitative analysis of the three opiates described in **Subheading 3.1.** and additionally for the analysis of cocaine and methadone (*see* **Fig. 2** for the structures of the compounds).

1. Before SFE , the hair is briefly washed by percolation with 10 mL of methylene chloride, 10 mL of water, and finally, 10 mL of methanol.
2. After this decontamination, drug abuser's hair is pulverized for 10 min with a ball-mill purchased from Retsch (Schieritz, Hauenstein, Switzerland).
3. Standard hair is prepared by adding an aqueous standard drugs solution to the pulverized hair. The mixture is submitted to magnetic stirring for 5 h, then filtered and the hair is washed with water and methanol.
4. The extraction cell is filled with pulverized hair (about 50 mg) and placed in the extraction chamber.
5. The oven is heated at 60°C and the restrictor is heated at 100°C.
6. The modifier pump is filled with the following modifier: methanol–water (4:1 v/v).
7. The hair samples are then extracted with 15% of the modifier in CO_2. The SFE conditions are the following: pressure: 350 atm, flow-rate: 1 mL/min, extraction time: 40 min or a 40-mL volume of extraction fluid.
8. During SFE, the drugs are trapped in 10 mL of methanol.
9. After extraction, the methanol is then evaporated to dryness under a gentle stream of nitrogen.
10. GC/MS is carried out with the same experimental conditions as in **Subheading 3.1.**: cocaine: m/z = 303 and 182, methadone: m/z = 294 and 72.

4. Notes

1. Choosing the right restrictor is a key factor in successful SFE. A major consideration is restrictor plugging caused by freezing during depressurization. Plugging is most pronounced with samples such as hair, having a high content of moisture, fat, or other aggregate-forming extractables. For all of these reasons, the use of a self-heated needle valve restrictor is necessary.
2. For quantitative analyses, nalorphine is recommended as a chromatographic standard or as an internal standard.
3. Since GC/MS represents the state-of-the-art for the identification and quantification of drugs in hair, the extracts obtained by these two methods could be analyzed by other techniques such as high performance liquid chromatography and capillary electrophoresis.

References

1. Baumgartner, A. M., Jones, P. F., Baumgartner, W. A., and Black, C. T. (1979) Radioimmunoassay of hair for determinig opiate-abuse histories. *J. Nucl. Med.* **20,** 749–752.
2. Walls, H. C. (1994) Drug testing in hair, a selective review of the literature. *Proceedings of the Society of Forensic Toxicologists Conference on Drug Testing in Hair, Tampa, FL.*
3. Kintz, P. (1996) Clinical applications of hair analysis, in *Drug Testing in Hair* (Kintz, P., ed.), CRC Press, Boca Raton, FL, pp. 267–277.
4. Sachs, H. (1996) Forensic applications of hair analysis, in *Drug Testing in Hair* (Kintz, P., ed.), CRC Press, Boca Raton, FL, pp. 211–222.
5. Tracqui, A. (1996) Le poil: structure et physiologie. *Rev. Fr. Labo.* **282,** 19–23.
6. Sachs, H. (1995) Theoretical limits of the evaluation of drug concentrations in hair due to irregular hair growth. *Forensic Sci. Int.* **70,** 53–61.
7. Kidwell, D. A. and Blank, D. L. (1996) Environmental exposure: the stumbling block of hair testing, in *Drug Testing in Hair* (Kintz, P., ed.), CRC Press, Boca Raton, FL, pp. 17–68.
8. Du Pont, R. L. and Baumgartner, W. A. (1995) Drug testing by urine and hair analysis: complementary features and scientific issues. *Forensic Sci. Int.* **70,** 63–76.
9. Kintz, P. (1996) Drug testing in addicts: a comparison between urine, sweat, saliva and hair. *Ther. Drug. Monit.* **18,** 450–455.
10. Harkey, M. R. (1993) Anatomy and physiology of hair. *Forensic Sci. Int.* **63,** 9–18.
11. Kintz, P. and Mangin, P. (1995) What constitutes a positive result in hair analysis: proposal for the establishment of cut-off values. *Forensic Sci. Int.* **70,** 3–11.
12. Moeller, M. R. (1992) Drug detection in hair by chromatographic procedures. *J. Chromatogr.* **580,** 125–134.
13. Kintz, P. (1993) Forensic hair examination: detection of drugs, in *Forensic Sci.* (Wecht, C., ed.), Matthew Bender, New York, pp. 1–32.
14. Inoue, T., Seta, S., and Goldberger, B. A. (1995) Analysis of drugs in unconventional samples, in *Handbook of Workplace Drug Testing* (Liu, R. H. and Golberger, B. A., eds.), AACC Press, Washington, D.C., pp. 131–158.
15. Moeller, M. R. and Eser, H. P. (1996) The analytical tools for hair testing, in *Drug Testing in Hair* (Kintz, P., ed.), CRC Press, Boca Raton, FL, pp. 95–120.
16. Cirimele, V., Kintz, P., Majdalani, R., and Mangin, P. (1995) Supercritical fluid extraction of drugs in drug addict hair. *J. Chromatogr. B.* **673,** 173–181.
17. Sachs, H. and Uhl, M. (1992) Opiat-Nachweis in Haar. Extrakten mit Hilfe von GC/MS/MS und Supercritical Fluid Extraction (SFE). *Toxichem. Krimtech.* **59,** 114–120.
18. Sachs, H. and Raff, I. (1993) Comparison of quantitative of drugs in human hair by GC/MS. *Forensic Sci. Int.* **63,** 207–216.
19. Edder, P., Staub, C., Veuthey, J. L., Pierroz, I., and Haerdi, W. (1994) Subcritical fluid extraction of opiates in hair of drug addicts. *J. Chromatogr. B.* **58,** 75–86.
20. Morrison, J. F., McCream, W. A., and Selavka, C. M. (1994) Evaluation of supercritical fluid extraction for the selective recovery of drugs of abuse from

hair. *Proceedings of the Second International Meeting on Clinical and Forensic Aspects of Hair Analysis, Genova, Italy, June 6–8.*

21. Staub, C., Edder, P., and Veuthey, J. L. (1996) Importance of supercritical fluid extraction (SFE) in hair analysis, in *Drug Testing in Hair* (Kintz, P., ed.), CRC Press, Boca Raton, FL, pp. 121–149.

13

Application of Direct Aqueous Supercritical Fluid Extraction for the Dynamic Recovery of Testosterone Liberated from the Enzymatic Hydrolysis of Testosterone-β-D-Glucuronide

Edward D. Ramsey, Brian Minty, and Anthony T. Rees

1. Introduction

The first report demonstrating the feasibility of supercritical fluids as solvent media for performing enzymatic reactions was published in 1986 *(1)*. Since then several reports have confirmed that the relatively low critical temperature and pressure of supercritical fluid carbon dioxide provides potential for the use of enzymes with thermally labile substrates. These applications, which generally involve the use of immobilized enzymes in feasibility studies for batch scale processes, have been reviewed *(2)*. One study *(3)* has demonstrated that the stability of nine commercially available enzyme preparations are largely unaffected using supercritical carbon dioxide containing ethanol (3–6%) and water (0.1%), at 35°C, 200 atm for 1 h. On the analytical scale, a few applications have described the use of enzymes in conjunction with supercritical fluid extraction (SFE) for sample preparation of liquid matrices *(4)*.

Testosterone is a naturally occurring male hormone whose administration can lead to artificial enhancement of strength and stamina. Dope testing methods at sport events involve the analysis of urine samples for surveillance purposes. Among these, gas chromatography combined with mass spectrometry (GC-MS) *(5,6)* and radioimmunoassay *(7)* procedures are used for the determination of abnormally high urinary testosterone levels. Since most of the testosterone is excreted in the form of glucuronic acid and sulfate conjugates that are too polar to be analyzed by GC-MS, methods *(5–7)* involve the incubation of urine samples with glucuronidase

From: *Methods in Biotechnology, Vol. 13: Supercritical Fluid Methods and Protocols*
Edited by: J. R. Williams and A. A. Clifford © Humana Press Inc., Totowa, NJ

for the liberation of testosterone before analysis. These procedures also require the use of solid-phase extraction (SPE) and liquid–liquid extraction preparation stages. For example, SPE can be used for the concentration and clean-up of testosterone (free and conjugated) prior to enzymatic hydrolysis *(5)* or for the isolation of free testosterone after enzymatic hydrolysis *(6)*.

This chapter provides a method whereby testosterone liberated from the enzymatic hydrolysis mixture of testosterone-β-D-glucuronide can be dynamically extracted using direct aqueous SFE with minimum sample handling. Analyte trapping is performed by decompressing the supercritical fluid extract onto an octadecyl silane (ODS) high performance liquid chromatography (HPLC) column. At the end of SFE, trapped testosterone is recovered from the HPLC column by solvent rinsing. The recovery of testosterone is finally determined using quantitative GC-MS with isotopically labeled testosterone acting as internal standard.

2. Materials

1. A 300-mL capacity direct aqueous SFE vessel (*see* **Note 1**). An official test certificate should certify that the vessel (embossed with a serial number) has been pressure-tested to 40 MPa—the minimum safe value for the experimental procedure described in **Subheading 3.**
2. A gas chromatographic oven, sufficiently large and strong enough to house the SFE vessel, e.g., a Pye 104 (Unicam, Cambridge, UK). The gas chromatograph should provide at least one port for the supply of HPLC grade 1/16 inch outer diameter inlet and outlet stainless-steel tubes to the SFE vessel.
3. A pumping system suitable for the delivery of 4 mL/min liquid carbon dioxide to the SFE vessel, e.g., a Gilson 307 reciprocating pump (Gilson, Villiers-le-Bel, France) fitted with a supercritical fluid grade piston seal and head cooling jacket.
4. A recirculator for passing coolant around the pump head assembly contained within the head cooling jacket, e.g., a Neslab RTE 110 recirculator (Neslab Instruments Inc., Newington, NH).
5. A programmable variable restrictor, fitted with supercritical fluid compatible seals, whose outlet is suitable for connection with high pressure HPLC compression fittings, e.g., Tescom Model 26-1722-24-084 (Tescom Corporation, Minneapolis, MN).
6. A cylinder of liquid carbon dioxide equipped with a liquid draw-off tube and on/off valve (*see* **Note 2**).
7. An HPLC pump capable of delivering 1 mL/min liquid, e.g., a second Gilson 307 pump.
8. A Gilson 811 mixing module or a HPLC 1/16 inch mixing “T” piece.
9. An HPLC column, 250 × 4.6 mm internal diameter, 5 μm ODS (*see* **Note 3**).
10. An HPLC six-port valve, e.g., a Rheodyne 7010 valve (Rheodyne, Cotati, CA).
11. A GC-MS instrument, e.g., a Hewlett-Packard 5971A MSD interfaced to a Hewlett-Packard 5890 gas chromatograph equipped with an HP-5MS 30 m × 0.3 mm

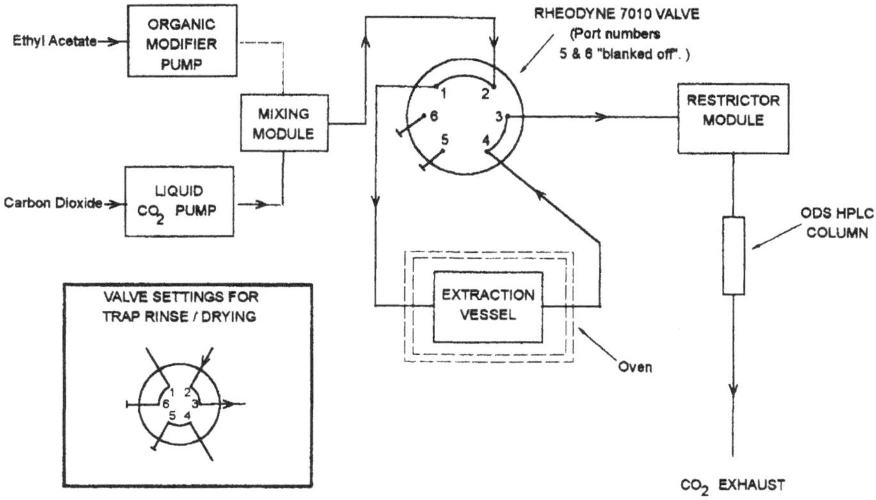

Fig. 1. Schematic of direct aqueous SFE system.

internal diameter column of 0.25 µm film thickness. Gas chromatography grade helium.

12. Class A, one mark borosilicate glass pipettes: 25 ± 0.03 mL and 50 ± 0.05 mL.
13. Class A, borosilicate glass volumetric flasks and stoppers: seven 100 ± 0.1 mL, three 500 ± 0.25 mL and one 10 ± 0.025 mL flasks.
14. A 5-mL continuously adjustable pipette with disposable tips, e.g., A Gilson Pipetman 5000.
15. Weighing balance, accurate to four decimal places. Reagent weighing boats.
16. Analar grade ethyl acetate, phosphoric acid, and deionized water.
17. Analytical grade samples of testosterone, 16,16,17-$^2H_3$-testosterone, and testosterone-β-D-glucuronide sodium salt.
18. β-Glucuronidase (*see* **Note 4**), type H-2: crude solution from *Helix pomatia* (β-glucuronidase activity: approx 100,000 U/mL at pH 5.0, 30-min assay).
19. Safety equipment: protective full-face screen, gloves.

3. Method

1. Construct the system from the components given above as shown in **Fig. 1**. All connections are made with 1/16 inch outer diameter HPLC stainless-steel tubing with appropriate compression fittings. The exhaust from the system should be vented into a fume hood to prevent any discharge of unretained testosterone into the laboratory atmosphere.
2. Prepare the SFE system for operation. Using a Gilson 307 pump in conjunction with the recirculator specified in the previous section, coolant at –10°C must be passed through the cooling jacket for at least 40 min before the pumping of liquid carbon dioxide. During this period, the SFE vessel should be equilibrated to 55°C.

3. Prime the HPLC pump with ethyl acetate.
4. Prepare the ODS HPLC column for analyte trapping. With the valve set to the trap rinse/drying position (shown in the inset of **Fig. 1**) and the restrictor module set to 0 MPa (i.e., fully open) flush the column using ethyl acetate at a flow rate of 1 mL/min for 10 min. Stop the flow of ethyl acetate and with the same valve and restrictor settings, dry the ODS column using carbon dioxide with a liquid flow rate of 3 mL/min for 10 min. At the end of this period, check the column gaseous exhaust by discharging onto a cooled surface to ensure the last traces of ethyl acetate have been displaced from the column. This having been achieved, set the liquid carbon dioxide flow rate to zero.
5. Prepare a fresh standard solution of testosterone-β-D-glucuronide sodium salt. Accurately weigh and transfer 3 mg of testosterone-β-D-glucuronide sodium salt into a 500-mL volumetric flask using a weighing boat. Dissolve and make up to 500 mL with deionized water.
6. Prepare a stock standard solution of testosterone. Accurately weigh and transfer 6 mg of testosterone into a 500 mL volumetric flask using a weighing boat. Dissolve and make up to 500 mL with Analar ethyl acetate. This will provide a testosterone solution of 12 ng/μL concentration.
7. Prepare a stock standard solution of 16,16,17-$^2H_3$-testosterone. Accurately weigh and transfer 4 mg of 16,16,17-$^2H_3$-testosterone into a 500 mL volumetric flask using a weighing boat. Dissolve and make up to 500 mL with Analar ethyl acetate. This will provide a 16,16,17-$^2H_3$-testosterone solution of 8 ng/μL concentration.
8. Prepare GC-MS calibration standards. Using the testosterone and 16,16,17-$^2H_3$-testosterone stock solutions, prepare five testosterone-ethyl acetate solutions whose concentrations of testosterone are: 6, 5, 3, 1.5 and 0.75 ng/μL. Each of the five testosterone solutions should also contain 16,16,17-$^2H_3$-testosterone at 2 ng/μL concentration. For example, to prepare a calibration standard whose testosterone and 16,16,17-$^2H_3$-testosterone concentrations are 6 and 2 ng/μL respectively: accurately transfer 50 mL of testosterone and 25 mL of 16,16,17-$^2H_3$-testosterone stock solutions (prepared in steps 6 and 7, respectively) into a 100 mL volumetric flask and make up with ethyl acetate.
9. Transfer 10 mL of the standard testosterone-β-D-glucuronide sodium salt solution to the direct aqueous SFE vessel followed by 220 mL dilute aqueous phosphoric acid solution (pH 5.2, prepared from the addition of Analar grade phosphoric acid to deionized water) and then 1.5 mL HP-2 β-glucuronidase.
10. Carry out direct aqueous SFE of the enzymatic digest. Check that (i) the SFE vessel is sealed, (ii) the restrictor module is set to deliver a back pressure of 24.1 MPa, and (iii) the valve is set to introduce carbon dioxide into the SFE vessel as shown in **Fig. 1**. Commence delivery of liquid carbon dioxide to the SFE vessel at 4 mL/min and ensure the exhaust from the ODS column is safely vented. After target pressure has been reached (approximately 15 min with the system described), perform dynamic SFE at 55°C for 120 min maintaining the liquid carbon dioxide flow rate at 4 mL/min.

11. Carry out the recovery of trapped testosterone from the ODS HPLC column. After 120 min dynamic SFE, the following sequence should be followed (i) switch the valve to the trap rinse/drying position, (ii) set the flow rate of liquid carbon dioxide to zero, and (iii) allow the ODS column to depressurize for 5 min with the restrictor now set to provide 0 MPa back pressure (fully open). Rinse the ODS column with 7 mL of ethyl acetate, using a flow rate of 1 mL/min, collecting the elute into a 10 mL volumetric flask. Stopper the flask and allow its contents to equilibrate to room temperature (*see* **Note 5**). Accurately add 2.5 mL of 16,16,17-$^2H_3$-testosterone stock solution (prepared in **step 7**) to the flask and finally make up to 10 mL with ethyl acetate.
12. Prepare the GC-MS system by tuning and calibrating the instrument.
13. Check GC-MS performance. The gas chromatograph temperature program used for all analyses is 2 min at 100°C, then to 290°C at 20°C/min, with the final temperature held for 10 min. The injection port and GC-MS interface temperatures should be set to 250°C and 300°C respectively. Inject 1 μL of testosterone stock solution (12 ng) in splitless mode and acquire full-scan electron ionization (EI) mass spectra, scanning the range 50–550 amu. The testosterone peak should be readily detected during a retention time window of 14–15.5 min. Using a Hewlett-Packard 5971A MSD, the EI mass spectrum obtained for testosterone should library search with a quality of fit typically greater than 95% (*see* **Note 6**).
14. Construct a GC-MS calibration graph by first creating a selected ion monitoring (SIM) data acquisition method. Monitor ions at m/z 288 and 246 for testosterone ($M_r = 288$) and m/z 291 and 249 for 16,16,17-$^2H_3$-testosterone ($M_r = 291$). Inject 3 μL of each of the five calibration standards (prepared in **step 8**) in splitless mode and analyze by SIM GC-MS. Use the quotient of responses obtained for the molecular ions of testosterone and 16,16,17-$^2H_3$-testosterone to construct a calibration graph with the ions at m/z 246 and 249 serving as qualifiers (*see* **Note 7**).
15. Determine the quantity of testosterone liberated by enzymatic hydrolysis of testosterone-β-D-glucuronide. Using splitless injection, analyze 3 μL of the 10 mL ethyl acetate solution prepared in **step 11** by SIM GC-MS. Once analysis has been performed, enable the GC-MS data system to use the previously constructed calibration graph to calculate the concentration of free testosterone by interpolation (*see* **Notes 8** and **9**).
16. Carefully discharge the contents of the SFE vessel to waste and rinse clean with deionized water.

4. Notes

1. The 300-mL capacity SFE vessel originally used for this application *(8)* was custom-built. Such vessels have to be manufactured to rigorous safety standards *(9,10)*. Alternatively, small volume (5–10 mL) commercially available direct aqueous SFE vessels can be used, scaling down the enzymatic hydrolysis with a reduced flow rate of liquid carbon dioxide. Suppliers of such vessels include Keystone Scientific (Bellefonte, PA) and Jasco Corporation (Tokyo, Japan).

2. The liquid carbon dioxide supply line from the cylinder leading to the pump inlet should be fitted with a filter assembly to prevent any particulates causing pump damage, e.g., a Nupro SS-TF 2 micron filter (Nupro Co., Willoughby, OH) is suitable for incorporation into 1/8 inch outer diameter stainless-steel supply lines.
3. If the ODS HPLC column serving as analyte trap has been newly shipped or has been used with aqueous mobile phases, the column should be first rinsed with ethanol prior to the conditioning stage described in the method section, **step 4**.
4. Only freshly supplied β-glucuronidase should be used. On receipt, the enzyme preparation should be immediately stored at 4°C with minimum exposure to light.
5. During SFE, the ODS column, which is not housed within a temperature regulated environment, undergoes rapid cooling at the point of decompression of the supercritical fluid. This can lead to ice formation on the outside of the column.
6. The Hewlett-Packard 5971A GC-MS data system automatically provides quality of fit values for EI library search results.
7. The GC-MS SIM method file described uses the option of qualifier ions. These ions are not used for quantitation but must be simultaneously detected with the ions that are used for quantitation. The ions at m/z 246 and 249 are detected at approx 40% relative abundance in the full-scan EI mass spectra of testosterone and 16,16,17-$^{2}H_{3}$-testosterone respectively.
8. Studies involving shorter or longer periods of dynamic aqueous SFE of the enzymatic hydrolysis can be performed. With the method described, approximately 70% of testosterone-β-D-glucuronide should have undergone enzymatic hydrolysis after 135 min (approximately 15 min are required to reach the target pressure of 24.1 MPa, during which the restrictor remains sealed, before the onset of 120 min dynamic SFE). Approximately 88% of the liberated testosterone should be trapped following 120 min dynamic SFE *(8)* using this procedure. N.B. These values are obtained providing less than 1 min separates the addition of β-glucuronidase and the delivery of liquid carbon dioxide to the SFE vessel (see method **steps 9** and **10**).
9. It has been demonstrated that in the absence of β-glucuronidase, testosterone-β-D-glucuronide is stable to hydrolysis using the direct aqueous SFE method *(8)*.

References

1. Nakamura, K., Chi, Y. M., Yamada, Y., and Yano, T. (1986) Lipase activity and stability in supercritical fluid carbon dioxide. *Chem. Eng. Commun.* **45,** 207–210.
2. Nakamura, K. (1990) Biochemical reactions in supercritical fluids. *TIBTECH* **8,** 288–292.
3. Taniguchi, M., Kamilhara, M., and Kobayashi, T. (1987) Effect of treatment with supercritical carbon dioxide on enzymatic activity. *Agric. Biol. Chem.* **51,** 593–594.
4. Ramsey, E. D., Minty, B., and Babecki, R. (1998) Supercritical fluid extraction strategies of liquid based matrices, in *Analytical Supercritical Fluid Extraction Techniques* (Ramsey, E. D., ed.) Kluwer Academic Publishers, Dordrecht, Netherlands, pp. 138–142.

5. Masse, R., Ayotte, C., and Dugal, R. (1989) Studies on anabolic steroids: integrated methodological approach to the gas chromatographic-mass spectrometric analysis of anabolic steroid metabolites in urine. *J. Chromatogr.* **489,** 23–50.
6. Houghton, E., Grainger, M. C., Dumasia, M. C., and Teale, P. (1992) Application of gas chromatography/mass spectrometry to steroid analysis in equine sports: problems with enzyme hydrolysis. *Organic Mass Spectrom.* **27,** 1061–1070.
7. Kicman, A. T., Brooks, R. V., Collyer, S. C., Cowan, D. A., Nanjee, M. N., Southan, G. J., and Wheeler, M. J. (1990) Criteria to indicate testosterone administration. *Br. J. Sports Med.* **24,** 253–264.
8. Ramsey, E. D., Minty, B., and Rees, A. T. (1996) Dynamic aqueous supercritical fluid extraction of the enzymic hydrolysis of testosterone-β-D-glucuronide. Analysis of liberated testosterone by gas chromatography-mass spectrometry. *Anal. Comm.* **33,** 307–309.
9. Saito, M. and Yamauchi, Y. (1994) Instrumentation, in *Fractionation by Packed-Column SFC and SFE* (Saito, M., Yamauchi, Y., and Okuyama, T., eds.), VCH Publishers, New York, pp. 101–133.
10. Taylor, L. T. (1996) *Supercritical Fluid Extraction.* Wiley, New York, pp. 53–57.

14

Analysis of Anabolic Drugs by Direct Aqueous Supercritical Fluid Extraction Coupled On-Line with High-Performance Liquid Chromatography

Edward D. Ramsey, Brian Minty, and Anthony T. Rees

1. Introduction

The analysis of drugs and their metabolites in biological fluids represents an essential role in pharmaceutical and toxicology studies. High performance liquid chromatography (HPLC) has emerged as a particularly powerful analytical technique for drug analysis since many water soluble compounds are too involatile and/or thermally labile to be analyzed by gas chromatography.

The on-line coupling of supercritical fluid extraction (SFE) with HPLC (SFE-HPLC) is technically challenging since the large volume of gas ultimately produced by SFE sample preparation is incompatible with HPLC operation, that is, possible admission of gas into the liquid mobile phase can lead to erratic HPLC pump and detector performance. Despite these problems, several reviews have described the use of SFE-HPLC *(1–3)*. Since many drugs are only present at ultra trace levels within liquid matrices, the development of appropriate SFE-HPLC methods offers considerable potential for the reduction of the number of sample handling stages with associated errors. Furthermore, direct aqueous SFE-HPLC is particularly well suited for the analysis of analytes which are light and/or air sensitive.

This chapter describes a relatively simple and convenient procedure whereby an SFE system equipped with a direct aqueous SFE vessel can be interfaced to HPLC instrumentation. The technique *(4)* uses a system of coupled octadecylsilane (ODS)-aminopropyl HPLC columns connected to the outlet of the SFE vessel. Moderately polar analytes which can be extracted are trapped onto the nonpolar ODS column during SFE. After SFE, these compounds are eluted

From: *Methods in Biotechnology, Vol. 13: Supercritical Fluid Methods and Protocols*
Edited by: J. R. Williams and A. A. Clifford © Humana Press Inc., Totowa, NJ

from the ODS column, with band focusing, onto the polar aminopropyl column using a nonpolar organic solvent. Finally, HPLC analysis is performed using a gradient program which slowly introduces a polar solvent into the mobile phase. The methodology is illustrated with reference to the analysis of anabolic drugs dissolved in water at the part-per-billion level using ultraviolet/visible (UV/VIS) diode array detection (DAD).

2. Materials

1. A 300-mL capacity direct aqueous SFE vessel (*see* **Note 1**). An official test certificate should certify that the vessel (embossed with a serial number) has been pressure tested to 40 MPa: the minimum safe value necessary for the experimental procedure described in **Subheading 3.**
2. A gas chromatographic oven, sufficiently large and strong enough to house the SFE vessel, e.g., a Pye 104 (Unicam, Cambridge, UK). The gas chromatographic oven should provide at least one port for the supply of HPLC grade 1/16" outer diameter inlet and outlet stainless-steel tubes to the SFE vessel.
3. A programmable variable restrictor, fitted with supercritical fluid compatible seals, whose outlet is suitable for connection with high-pressure HPLC compression fittings, e.g., Tescom Model 26-1722-24-084 (Tescom Corporation, Minneapolis, MN).
4. A pumping system suitable for the delivery of liquid carbon dioxide to the SFE vessel, e.g., a Gilson 307 reciprocating pump (Gilson, Villiers-le-Bel, France) fitted with a supercritical fluid chromatography grade piston seal and head cooling jacket.
5. A recirculator for passing coolant around the pump head assembly contained within the head cooling jacket, e.g., a Neslab RTE 110 recirculator (Neslab Instruments Inc., Newington, NH).
6. A cylinder of liquid carbon dioxide equipped with a liquid draw-off tube and on/off valve (*see* **Note 2**).
7. A gradient HPLC system equipped with UV/VIS DAD facilities and six-port variable loop sample injection valve.
8. A high-pressure 10-port switching valve, e.g., Rheodyne 7610-400 (Cotati, CA) or a Valco C2-2000 (Houston, TX).
9. Two Valco 1/16 inch low dead volume stainless-steel unions, equipped with 6000 psi (rated for stainless-steel tubing) one-piece fingertight polymeric fittings. An appropriate length of 1/16 inch outer diameter HPLC grade stainless-steel tubing (*see* **Subheading 3., steps 1** and **7**) fitted with one-piece 6000 psi-rated fingertight fittings.
10. HPLC grade heptane, ethanol, and deionized water.
11. Two HPLC columns: (i) a 150 × 4.6 mm internal diameter, 5 μm aminopropyl column, and (ii) a 250 × 4.6 mm internal diameter, 5 μm ODS column (*see* **Note 3**).
12. Analytical grade samples of estrone and zeranol.
13. Volumetric flasks (10 mL and 250 mL), weighing balance accurate to four decimal places. A 1-mL adjustable pipet, e.g., Gilson P1000, which can accurately dispense 100 μL.
14. Safety equipment: protective full-face screen, gloves.

3. Method

1. Assemble **items 1–3** in the Materials section. The SFE vessel is installed in the oven and provided with inlet and outlet connections consisting of 1/16 inch outer diameter HPLC grade stainless-steel tubing. The restrictor is then connected to the outlet.
2. Assemble **items 4–6** in **Subheading 2.** to provide the supply of liquid carbon dioxide. The recirculator is fitted to supply cooling fluid to the pump head and the cylinder of carbon dioxide is connected to the pump inlet.
3. Then assemble the entire system as shown in **Fig. 1**, which also shows the two valve configurations used during operation (*see* **Note 4**). All connections shown are made with 1/16 inch outer diameter HPLC grade stainless-steel tubing. The exhaust from the system should be led to a fume hood when drug solutions are being extracted.
4. Prepare the SFE system. Using a Gilson 307 pump in conjunction with the recirculator specified in the previous section, coolant at –10°C must be passed through the cooling jacket for at least 40 min before the pumping of liquid carbon dioxide. During this period, the direct aqueous SFE vessel should be equilibrated to 55°C.
5. Prepare the HPLC system. Prime the pumps to deliver heptane and ethanol, respectively. Set the DAD to scan through the range 200–500 nm.
6. Carry out stage 1 of the HPLC columns conditioning cycle. Select the valve setting shown in **Fig. 1(B)**. Condition the coupled ODS-aminopropyl columns using pure heptane at 1 mL/min for 15 min. During this period check DAD stability. At the end of the first stage of column conditioning, set the HPLC flow to zero and allow the columns to depressurize.
7. Carry out stage 2 of the HPLC columns conditioning cycle. Switch the ten 10-port valve to the position shown in **Fig. 1(A)**. Bypass the restrictor module and SFE vessel by linking the outlet from union B to the inlet of union A. This can be achieved using a length of 1/16 inch OD HPLC stainless-steel tubing fitted with one-piece 6000 psi-rated fingertight fittings at each end. Set the SFE pump to deliver a liquid carbon dioxide flow rate of 2 mL/min for 10 min to dry the ODS column. At the end of this period, inspect the column gaseous exhaust, discharging onto a cooled surface, to ensure the last traces of heptane have been displaced from the ODS column. Once this has been achieved, set the flow of liquid carbon dioxide to zero and wait until the flow of gaseous carbon dioxide from the column has ceased. Reconnect the restrictor module and SFE vessel into the system by means of the appropriate union plumbing. The ODS column should now be dry with the aminopropyl column primed with heptane.
8. Preparation of estrone and zeranol standard solution. Dissolve 5 mg of each compound in approximately 8 mL ethanol within a 10 mL volumetric flask. After complete dissolution, make-up to 10 mL.
9. Load the SFE vessel with 250 mL of water spiked with estrone and zeranol, each at the 200 ppb level, i.e., 100 µL of the drug solution made in **step 8** is added to 250 mL water.
10. Direct aqueous SFE. Check that (i) the SFE vessel is sealed, (ii) the valve configuration is set to the position shown in **Fig. 1(A)**, (iii) the restrictor module is

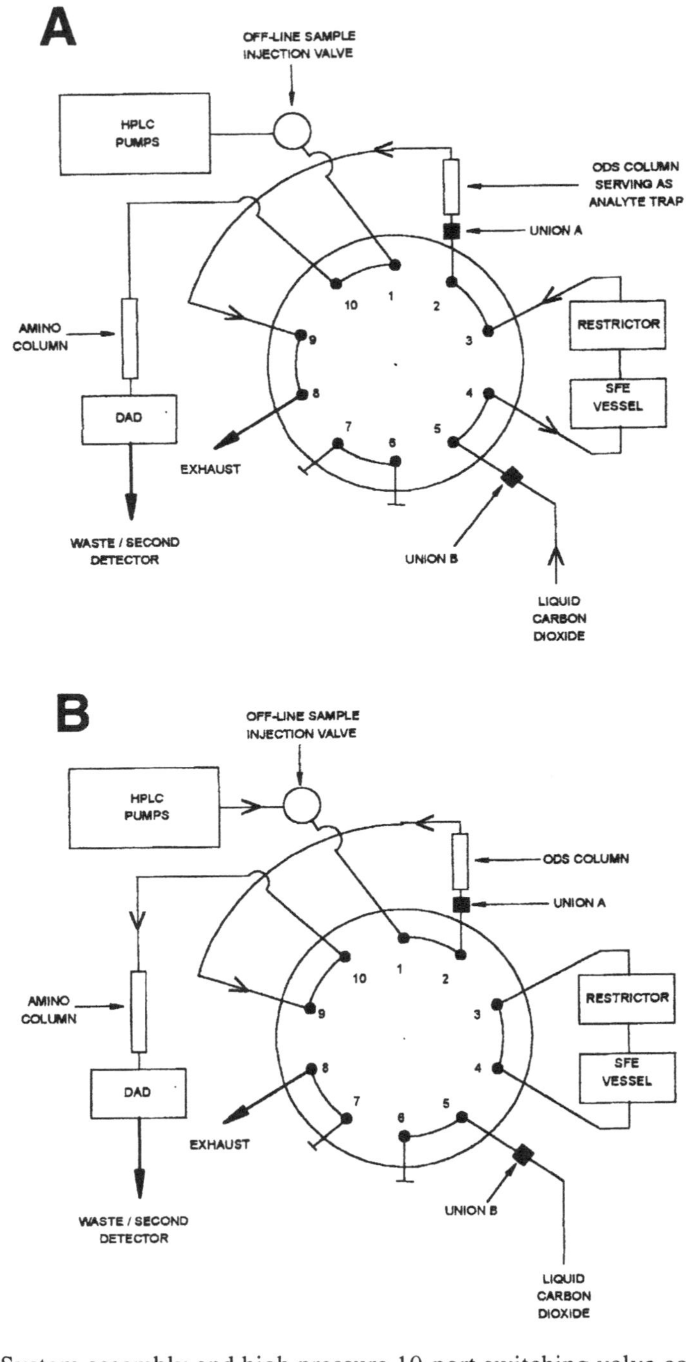

Fig. 1. System assembly and high pressure 10-port switching valve configurations for (**A**) SFE and (**B**) SFE-HPLC.

programmed to deliver a back pressure of 24.1 MPa, and (iv) the exhaust from the ODS column is safely vented. Perform direct aqueous SFE at 55°C for 30 min with a liquid carbon dioxide flow rate of 8 mL/min.

11. The following sequence should be followed between SFE and HPLC analysis: (i) set the flow rate of liquid carbon dioxide to zero, (ii) check the restrictor has sealed, and (iii) allow the ODS column to depressurize for 5 min (*see* **Note 5**).
12. Carry out the HPLC analysis of extracted estrone and zeranol (*see* **Note 6**). Switch the valve to the position shown in **Fig. 1(B)**. Analysis is performed using pure heptane for 12 min then to heptane-ethanol (65 + 35, parts volume) at time 42 min following a linear profile at flow rate 1 mL/min (*see* **Notes 7** and **8**). Selection of the DAD two-dimensional chromatogram at 281 nm facilitates monitoring the elution of the example drugs (*see* **Note 9**).
13. Carefully vent the direct aqueous SFE vessel, empty and rinse clean with deionized water.

4. Notes

1. The 300 mL capacity SFE vessel originally used for this application *(4)* was custom-built. Such vessels have to be manufactured to rigorous safety standards *(5,6)*. Alternatively, small volume (5–10 mL) commercially available direct aqueous SFE vessels can be used for this SFE-HPLC procedure, using a reduced flow rate of liquid carbon dioxide. Suppliers of such vessels include Keystone Scientific (Bellefonte, PA) and Jasco Corp. (Tokyo, Japan).
2. The liquid carbon dioxide supply line from the cylinder leading to the pump inlet should be fitted with a filter assembly to prevent any particulates causing pump damage, e.g., a Nupro SS-TF 2 µm filter (Nupro Co., Willoughby, OH) is suitable for incorporation into 1/8 inch outer diameter stainless-steel supply lines.
3. If new HPLC columns are used which have been shipped containing methanol or the presence of water is suspected, the columns should be first conditioned with pure ethanol. Methanol and heptane are immiscible.
4. By incorporating an additional high pressure switching valve (e.g., Rheodyne 7010) into the system, the need to undo plumbing to bypass the restrictor module and SFE vessel whilst the ODS column is dried can be avoided. *See* Chapter 13, **Fig. 1**. With this arrangement, the two low dead volume 1/16 inch unions can be eliminated.
5. During SFE, the ODS column, which is not housed within a temperature regulated environment, undergoes rapid cooling at the point of decompression of the supercritical fluid. This can lead to ice formation on the outside of the column.
6. The retention times of estrone and zeranol can be determined using the valve configuration shown in **Fig. 1(B)** without using an SFE stage. Sample introduction onto the coupled columns (both primed with heptane) can be made using the off-line variable loop HPLC injection valve. After injection, the gradient elution program can be run in the normal manner. Also off-line HPLC analyses, using only the aminopropyl column, can be performed with the valve configuration shown in **Fig. 1(A)**.
7. The normal phase gradient program uses an initial step of pure heptane for a period of 12 min. This stage ensures trapped drugs are eluted from the ODS col-

umn onto the amino column and that any residual carbon dioxide is purged from the system before the onset of chromatography.

8. The band focusing effect of the coupled column system is such *(4)* that an internal standard (of suitable polarity) can be injected post-SFE using the off-line sample injection valve during the initial 12-min isocratic period. This capability helps facilitate quantitative SFE-HPLC studies.
9. A second detector, e.g., a mass spectrometer equipped with an appropriate liquid chromatography interface can be connected in series *(4)* with the diode array detector.

References

1. Howard, A. L. and Taylor, L. T. (1993) Supercritical fluid extraction-high performance liquid chromatography: on-line and off-line strategies, in *Supercritical Fluid Extraction and Its Use in Chromatographic Sample Preparation* (Westwood, S. A., ed.) Blackie Academic and Professional, London, pp. 145–168.
2. Griebrokk, T. (1995) Applications of supercritical fluid extraction in multidimensional systems. *J. Chromatogr. A* **703,** 523–536.
3. Rees, A. T. (1998) Supercritical fluid extraction for off-line and on-line high performance liquid chromatographic analysis, in *Analytical Supercritical Fluid Extraction Techniques* (Ramsey, E. D., ed.) Kluwer Academic Publishers, Dordrecht, Netherlands, pp. 330–349.
4. Ramsey, E. D., Minty, B., and Rees, A. T. (1997) Drugs in water: analysis at the part-per-billion level using direct supercritical fluid extraction of aqueous samples coupled on-line with ultraviolet-visible diode-array liquid chromatography-mass spectrometry. *Anal. Comm.* **34,** 51–54.
5. Saito, M. and Yamauchi, Y. (1994) Instrumentation, in *Fractionation by Packed-Column SFC and SFE* (Saito, M., Yamauchi, Y., and Okuyama, T., eds.), VCH Publishers Inc., New York, pp. 101–133.
6. Taylor, L. T. (1996) *Supercritical Fluid Extraction.* Wiley, New York, pp. 53–57.

15

Detection of Beta-Blockers in Urine and Serum by Solid-Phase Extraction–Supercritical Fluid Extraction and Gas Chromatography–Mass Spectrometry

Kari Hartonen and Marja-Liisa Riekkola

1. Introduction

Beta-blockers have some clinical use in the treatment of angina pectoris, hypertension, and tachycardia. In addition, they have been used to control migraine, chronic alcoholism, schizophrenia, essential tremor, and cardiac effects of cocaine overdose. Unfortunately, misuse of these drugs as doping agents in archery, billiards, and riflery competitions happens from time to time where 5- to 100-mg oral doses are typical to decrease the heart rate and muscular tremor *(1,2)*.

Determination of β-blockers in biological fluids like urine and serum by liquid chromatography (LC) *(3)*, capillary electrophoresis *(4)*, or gas chromatography (GC) *(5,6)* can be difficult due to their low concentrations relative to the high concentration of endogenous compounds in the sample matrix. Since β-blockers are also metabolized in a matter of hours, and they appear in varying hydrophilicity and protein binding capabilities *(7)*, their determination is even more complicated. Several extraction and clean-up methods for β-blockers have been applied *(8,9)*, including liquid–liquid extraction (LLE) and solid-phase extraction (SPE). Recently, an on-line combination of reversed phase LC with GC using on-line LLE has been successfully applied to determine β-blockers in urine and serum *(10)*. SPE has proven to be superior over LLE, giving good recovery for hydrophilic and hydrophobic compounds at the same time *(9,11)*. Both methods usually need a separate deproteinization step and derivatization if GC is used for analysis. Sometimes with LLE it is necessary to back-extract the sample into the aqueous phase to clean it up.

From: *Methods in Biotechnology, Vol. 13: Supercritical Fluid Methods and Protocols*
Edited by: J. R. Williams and A. A. Clifford © Humana Press Inc., Totowa, NJ

Owing to the special properties of supercritical fluids, supercritical fluid extraction (SFE) is much faster and usually more efficient than conventional solvent extraction methods (*see* Chapter 1). Using the most common fluid carbon dioxide for the extraction will result in clean extracts with a few interferences, and extracts can also be obtained in highly concentrated form because of the volatility of the fluid. SFE works well for organics in solid sample matrices such as soil ***(12)***, sediment ***(13)***, flyash ***(14)***, various sorbent materials ***(15)***, and food products ***(16)***. Several biologically and pharmaceutically interesting SFE reports can also be found ***(17–21)***. SFE is a serious alternative to many Soxhlet and sonication-assisted extraction methods ***(22,23)***.

Direct SFE of liquid (aqueous) sample matrices is less frequently reported. This is probably due to the easiness of the sample matrix being flushed out with analytes and CO_2 if an unsuitable combination of sample volume and flow-rate (pressure), relative to the size of extraction vessel, is used (even with the extraction vessels developed for liquid sample matrices). In addition, the solubility of water in supercritical CO_2 is about 0.1%, which might result in blocking of the restrictor due to the freezing of water. These difficulties can be overcome by using special phase separators ***(24)***. However, with aqueous samples, the extraction temperature is very limited, and the extraction of polar and hydrophilic compounds with nonpolar CO_2 cannot be done successfully, since the use of polar modifiers is ruled out. Additionally, many of these polar analytes will need to be derivatized if GC is used to analyze them. In many cases, this cannot be done successfully, since water inhibits the derivatization reaction.

A very convenient way to process aqueous samples is the combination of SPE and SFE ***(6,25,26)***. This combines the selectivity of both methods and gives increased sample clean-up and fractionation capabilities. Cartridge or Sep-Pak-type SPE tubes can be inserted inside the SFE extraction vessel to elute analytes from the sorbent with the supercritical fluid instead of with an organic solvent ***(25)***. This can, however, produce some contamination originating from the plastic tubes, especially if modifiers are used or derivatization reagents are added to the extraction vessel (use of the tubes might then be completely impossible). Removing the sorbent quantitatively from the tube and transferring it into the extraction vessel is also quite tedious. A much better choice is to use SPE discs (Empore™ discs) ***(6)*** where sorbent is mixed with Teflon, thus producing a more inert tool to be used for the determination of very small amounts of organics. Additionally, discs are very flexible and can be more easily inserted inside the SFE extraction vessel. Larger sample loads and flow rates can also be used with these large-diameter discs compared to SPE tubes.

In this chapter, an effective protocol is described, where SFE is used to extract β-blockers from Empore™ C_{18} solid-phase extraction discs, which are used to collect those drugs from urine or serum samples. The number of pretreatment steps are minimized by derivatizing β-blockers in the SFE extraction vessel. Acetic anhydride is used as the derivatization reagent. After SPE-SFE, the acetylated drugs are analyzed by GC-MS.

2. Materials

1. Solid phase C_{18} discs (Empore™) with an outer diameter of 47 mm.
2. HPLC-grade methanol and methylene chloride.
3. Buffer solution, pH 10, of 0.01 M borax and 0.1 M NaOH.
4. Distilled and preferably deionized water.
5. A normal vacuum filtering device for 47 mm outer diameter filters.
6. SFE/SFC-grade or similar high purity carbon dioxide.
7. Analytical-grade acetic anhydride and pyridine for the derivatization. Purity of the reagents is critical for the acetylation to be successful, and both must be distilled before use, especially if they have been stored long time and are slightly colored. Store the reagents in a refrigerator.
8. SFE apparatus.
9. 5-mL extraction vessel in SFE.
10. 10-cm linear fused silica capillary with an inner diameter of 30 µm for use as a restrictor. This should be heated intrinsically or a hot air blower provided for restrictor heating.
11. A supply of 7.5 mL screw-cap glass vials, 17 mm × 61 mm.
12. A water bath or metal block heater for maintaining the collection vial at 5°C.
13. Nitrogen gas for evaporation of trapping solvent.
14. A gas chromatograph.
15. HP-5 or DB-5 type GC column with a minimum length of 15 m and with an inner diameter of 0.2 mm.
16. A 2.5 m long retention gap with 0.32 or 0.53 mm inner diameter depending on the outer diameter of the syringe needle for on-column injection. For autosamplers, a 0.53 mm inner diameter retention gap must be used. The retention gap is connected to the analytical column with glass pressfit connector.
17. A mass spectrometer for use with El (electron impact) ionization and in SIM mode.
18. Helium for use as a carrier gas in GC.

3. Method

3.1. Preparation of Solutions for Calibration

1. Prepare stock solution (A) of standard β-blockers to be determined in methanol (5 mg/mL) and a separate solution (B) of a β-blocker or other similar compound (5 mg/mL) in methanol to be used as an internal standard. Store the solutions in a refrigerator.
2. Prepare standard solutions (S) for linear calibration plots by taking 10, 25, 50, 100, 150, and 200 µL of solution A and add 150 µL of solution B to each and dilute these to 5 mL with methanol.

3. Make an additional solution (C) for the internal standard by taking 150 μL of solution B and diluting with methanol to 5 mL.

3.2. Sample Pretreatment

1. Store blank and real urine and serum samples in the freezer (*see* **Note 1**).
2. Let the urine and serum samples melt and warm up to room temperature. Shake each sample before taking an aliquot for analysis to ensure homogeneity of the sample.
3. Spike the blank urine samples (2 mL) or the blank serum samples (1 mL) with 50 μL of the already prepared standard solutions (S) and after dilution with the buffer, process them in the same way as the real samples by SPE-SFE and GC-MS. These calibration samples correspond to β-blocker concentrations of 0.25, 0.63, 1.25, 2.5, 3.75, and 5 μg/mL of urine, if 2 mL urine are used. Scale can be extended depending on the analyte concentrations in the samples, the sample volume and the sensitivity of the MS. Use the same amount of blank urine (or serum) as the actual sample size.
4. Take 2 mL of real urine or 1 mL of real serum sample and add it to the test tube. Sample size can be increased to 5 mL with urine and to 3 mL with serum if the level of analytes is very low. Add 50 μL of internal standard solution C to each sample.
5. Dilute the urine sample 2:3 (v/v) and the serum sample 1:5 (v/v) with borax buffer adjusted to pH 10 with NaOH. Do the same also for the calibration samples.

3.3. Solid-Phase Extraction

1. Place the C_{18} disc in to the filtering device and apply the vacuum. Wash the disc with 20 mL of methanol, 10–20 mL of water, and with 10 mL of buffer solution to adjust the pH of the sample. Care should be taken not to let the disc run dry between and after these steps (*see* **Note 2**).
2. Introduce the sample on to the disc and use 2 × 5 mL of buffer solution to wash the sample tube and filtering system. The sample and washing buffer should be slowly filtered through the disc with the vacuum at 1–2 mL/min.
3. Dry the disc with full vacuum for at least 7 min.
4. Transfer the disc into the SFE extraction vessel.

3.4. Supercritical Fluid Extraction

1. Place the disc containing the analytes in the extraction vessel, close the bottom end of the extraction vessel and connect this to the inlet of CO_2. The extraction vessel should be placed vertical with the flow of CO_2 upward (*see* **Note 3**).
2. Add 150–200 μL of acetic anhydride and 400 μL of pyridine to the extraction vessel. Use glass pipettes rather than plastic ones.
3. Close the upper end of the vessel and connect it to the outlet capillary leading to the pressure restrictor.
4. Connect a new silica restrictor to the exit of the system and place the exit end of the restrictor into the collection vial containing 3.5–4 mL of methanol (*see* **Note 4**).

5. Keep the collection vial thermostatted at ca. +5°C (*see* **Note 5**).
6. Set the extraction pressure to 400–500 atm (depending on the instrument maximum) and temperature to 120–150°C (*see* **Note 6**).
7. Use a 1- to 5-min static extraction at the beginning to let the acetylation occur. Depending on the SFE instrument, it will take time to reach the desired temperature (equilibrium time), which usually is enough for the acetylation reaction. This is why the static extraction period can be short (*see* **Note 7**).
8. After static SFE, extract the acetylated β-blockers dynamically using 40 g of CO_2 per extraction. Make sure that the CO_2 flow rate remains approximately constant. Flow rate under these conditions, and with the restrictor as described earlier, should be about 1.5 mL/min (measured at the pump). Heating the restrictor will help if the flow rate is decreasing due to a partially blocked restrictor (*see* **Note 8**).
9. Because of the gas flow through the collection solvent, some methanol will be evaporated, and more has to be added from time to time to keep the solvent level constant.
10. After completing the extraction, evaporate the extract to dryness under nitrogen flow at 60–80°C. Redissolve the sample in 200 μL of methylene chloride:methanol (9:1, v/v) for analysis by GC-MS (*see* **Note 9**).
11. Clean the extraction vessel in an ultrasonic bath between extractions. Use methanol as a cleaning solvent.

3.5. Gas Chromatography–Mass Spectrometry

1. Inject 1–2 μL of sample (with an on-column technique) directly onto the column.
2. If an autosampler is used, insert tubes for the sample vials are necessary to get the sample into the syringe.
3. Perform all injections at 30°C and at constant pressure (60 kPa). A suitable temperature program for the GC oven when using a 15-m column is from 30°C (2 min) to 220°C at 15°C/min, from 220°C (1 min) to 260°C at 5°C/min, and from 260°C to 320°C (3 min) at 15°C/min.
4. In SIM mode, record the ions at m/z 72, 158, and 200, and use the subtracted ion (at m/z 200) chromatogram for quantitation. In addition to retention time, use relative intensities of all the ions for identification (qualification).
5. Run the calibration standards first (starting from the most diluted one) and make a calibration curve for each β-blocker (separate curves for urine and for serum).
6. After the calibration standards, run the actual urine and serum samples (*see* **Note 10**).

4. Notes

1. The amount of the metabolites will increase as a function of time between the intake of the β-blockers and the collection of urine or serum samples. This will make the detection of β-blockers more difficult and screening of the amount of metabolites more important.
2. In SPE, the sorbent must be conditioned. Reverse phase sorbents are usually very hydrophobic, and they need some organic solvent to solvate or wet their surfaces.

Without this layer of organic solvent, poor extraction and difficulties with passing water through the sorbent may occur.

3. Pushing the reagents into the tubings after the extraction vessel during the pressurization can be avoided (minimized) using an upward flow during SFE.
4. New commercial SFE instruments are equipped with an automated and adjustable restrictor and a 1/16-inch stainless steel tube can be connected to the outlet of the restrictor and the exit of the tube immersed into the collection solvent. The tube must be eluted with 2–3 mL of methanol after the extraction and the effluent combined with the extract.
5. If the collection solvent remains about +5°C, efficient collection will occur *(27)*. Of course, with long extraction times (>30 min), the water in the bath or the metal block may start to cool and the temperature of the solvent will decrease.
6. Slightly increased extraction efficiency might be obtained at 150°C than at 120°C, but the lifetime of the seals in the extraction vessel is then greatly decreased with a greater risk of leaking.
7. In SFE, the acetylation reaction is very fast and with longer static extraction times, a decrease in recovery has been noticed *(28)*.
8. A hot-air gun will be fine for restrictor heating, but heating the collection solvent should be avoided (to minimize the evaporation of the collection solvent).
9. Even a small amount of acetylation reagents will have a dramatic effect on chromatographic separation. Methanol is needed because pure methylene chloride does not always dissolve all the β-blockers.
10. When chromatographic peaks start to tail, or are otherwise bad, about 30 cm can be cut away from the beginning of the retention gap to restore good peak shapes. This can be done only two or three times before changing to a completely new retention gap.

References

1. Park, J., Park, S., Lho, D., Choo, H. P., Chung, B., Yoon, C., Min, H., and Choi, M. J. (1990) Drug testing at the 10th Asian games and 24th Seoul Olympic games. *J. Anal. Toxicol.* **14,** 66.
2. Leloux, M. S. and Dost, F. (1991) Doping analysis of beta-blocking drugs using high-performance liquid chromatography. *Chromatographia* **32,** 429.
3. Ahnoff, M., Ervik, M., Lagerstrom, P.-O., Persson, B.-A., and Vessman, J. (1985) Drug level monitoring: cardiovascular drugs. *J. Chromatogr.* **340,** 73.
4. Lukkari, P., Sirén, H., Pantsar, M., and Riekkola, M.-L. (1993) Determination of ten β-blockers in urine by micellar electrokinetic capillary chromatography. *J. Chromatogr.* **632,** 143.
5. Lho, D.-S., Hong, J.-K., Paek, H.-K., Lee, J.-A., and Park, J. (1990) Determination of phenolalkylamines, narcotic analgesics, and beta-blockers by gas chromatography/mass spectrometry. *J. Anal. Toxicol.* **14,** 77.
6. Hartonen, K. and Riekkola, M.-L. (1996) Detection of β-blockers in urine by solid-phase extraction-supercritical fluid extraction and gas chromatography-mass spectrometry. *J Chromatogr. B* **676,** 45.

7. Marko, V. (1989) Determination of beta-blockers in biological material, in *Techniques and Instrumentation in Analytical Chemistry,* Vol. 4, Part C, Elsevier, Amsterdam, ch. 1 and 3.
8. Sirén, H., Saarinen, M., Hainari, S., Lukkari, P., and Riekkola, M.-L. (1993) Screening of beta-blockers in human serum by ion-pair chromatography and their identification as methyl or acetyl derivatives by gas chromatography-mass spectrometry. *J. Chromatogr.* **632,** 215.
9. McDowall, R. D., Pearce, J. C., and Murkitt, G. S. (1986) Liquid-solid sample preparation in drug analysis. *J. Pharm. Biomed Anal.* **4,** 3.
10. Hyötyläinen, T., Andersson, T., and Riekkola, M.-L. (1997) Liquid chromatographic sample cleanup coupled on-line with gas chromatography in the analysis of beta-blockers in human serum and urine. *J. Chromatogr. Sci.* **35,** 280.
11. Leloux, M. S., DeJong, E. G., and Maes, R. A. A. (1989) Improved screening method for beta-blockers in urine using solid-phase extraction and capillary gas chromatography-mass spectrometry. *J. Chromatogr.* **488,** 357.
12. Wenclawiak, B., Rathmann, C., and Teuber, A. (1992) Supercritical fluid extraction of soil samples and determination of polycyclic aromatic hydrocarbons (PAHs) by HPLC. *Fresenius J. Anal. Chem.* **344,** 497.
13. Meyer, A. and Kleiböhmer, W. (1993) Supercritical fluid extraction of polycyclic aromatic hydrocarbons from a marine sediment and analyte collection via liquid-solid trapping. *J. Chromatogr. A* **657,** 327.
14. Hills, J. W., Hill, H. H., Hansen, D. R., and Metcalf, S. G. (1994) Carbon dioxide supercritical fluid extraction of incinerator fly ash with a reactive solvent modifier. *J. Chromatogr. A* **679,** 319.
15. Hawthorne, S. B., Krieger, M. S., and Miller, D. J. (1989) Supercritical carbon dioxide extraction of polychlorinated biphenyls, polycyclic aromatic hydrocarbons, heteroatom-containing polycyclic aromatic hydrocarbons, and *n*-alkanes from polyurethane foam sorbents. *Anal. Chem.* **61,** 736.
16. King, J. W., Johnson, J. H., and Friedrich, J. P. (1989) Extraction of fat tissue from meat products with supercritical carbon dioxide. *J. Agr. Food Chem.* **37,** 951.
17. Edder, P., Veuthey, J. L., Kohler, M., Staub, C., and Haerdi, W. (1994) Subcritical fluid extraction of morphinic alkaloids in urine and other liquid matrices after adsorption on solid supports. *Chromatographia* **38,** 35.
18. Phillips, E. M. and Stella, V. J. (1993) Rapid expansion from supercritical solutions: application to pharmaceutical processes. *Int. J. Pharm.* **94,** 1.
19. Ramsey, E. D., Perkins, J. R., Games, D. E., and Startin, J. R. (1989) Analysis of drug residues in tissue by combined supercritical fluid extraction-supercritical fluid chromatography-mass spectrometry-mass spectrometry. *J. Chromatogr.* **464,** 353.
20. Messer, D. C., Taylor, L. T., Moore, W. N., and Weiser, W. E. (1993) Assessment of supercritical fluids for drug analysis. *Ther. Drug Monit.* **15,** 581.
21. Moore, W. N. and Taylor, L. T. (1994) Analytical inverse supercritical fluid extraction of polar pharmaceutical compounds from cream and ointment matrices. *J. Pharm. Biomed Anal.* **12,** 1227.

22. Onuska, F. I., Terry, K. A., and Wilkinson, R. J. (1993) The analysis of chlorinated dibenzofurans in municipal fly ash: supercritical fluid extraction vs Soxhlet. *High Res. Chromatogr.* **16,** 407.
23. Richards, M. and Campbell, R. M. (1991) Comparison of supercritical fluid extraction, Soxhlet, and sonication methods for the determination of priority pollutants in soil. *LC-GC Int.* **4,** 33.
24. Thiebaut, D., Chervet, J.-P., Vannoort, R. W., DeJong, G. J., Brinkman, U. A. Th., and Frei, R. W. (1989) Supercritical fluid extraction of aqueous samples and on-line coupling to supercritical fluid chromatography. *J. Chromatogr.* **477,** 151.
25. Liu, H. and Weluneyer, K. R. (1992) Solid-phase extraction with supercritical fluid elution as a sample preparation technique for the ultratrace analysis of flavone in blood plasma. *J. Chromatogr. B* **577,** 61.
26. Tang, P. H.-T. and Ho, J. S. (1994) Liquid-solid disk extraction followed by supercritical fluid elution and gas chromatography of phenols from water. *High Res. Chromatogr.* **17,** 509.
27. Langenfeld, J. J., Burford, M. D., Hawthorne, S. B., and Miller, D. J. (1992) Effects of collection solvent parameters and extraction cell geometry on supercritical fluid extraction efficiencies. *J. Chromatogr.* **594,** 297.
28. Meissner, G., Hartonen, K., and Riekkola, M.-L. (1998) Supercritical fluid extraction combined with solid-phase extraction as sample preparation technique for the analysis of β-blockers in serum and urine. *Fresenius J. Anal. Chem.* **360,** 618.

16

On-Line SFE-SFC for the Analysis of Fat-Soluble Vitamins and Other Lipids from Water Matrices

Francisco J. Señoráns and Karin E. Markides

1. Introduction

The determination of trace organic compounds in aqueous samples usually involves isolation of the fraction of interest followed by subsequent separation by means of a chromatographic technique. When high resolution is needed, the main chromatographic techniques usually employed are capillary gas chromatography (GC) or supercritical fluid chromatography (SFC), both as analytical tools in themselves and as inlet methods for mass spectrometry (MS) *(1)*. SFC (*see* Chapter 1) has features overlapping gas and liquid chromatography, and may use numerous detectors under mild conditions, including the universal flame ionization detector (FID) and improved chromatographic–mass spectrometric interfaces, that opens additional possibilities for the study of retinoids, carotenoids, other vitamins, and related compounds *(2)*. One drawback of the SFC techniques when using carbon dioxide as mobile phase is that the direct introduction of water samples poses a series of problems. Water must therefore be eliminated before it reaches the analytical column. A sample preparation step is thus essential to both concentrate the sample and eliminate the water.

This sample pretreatment may be carried out in different ways, mainly liquid–liquid extraction, solid-phase extraction (SPE), and supercritical fluid extraction (SFE). Sample preparation by supercritical fluid extraction has recently had a rapid expansion of applications and demonstrated to have a number of advantages compared to traditional methods, including shorter extraction times, tunable selectivity (i.e., selective extractions of analytes by varying the pressure or temperature) and organic solvent use minimization *(3)*. Supercritical carbon dioxide in particular is of especial interest to the biochemical laboratories and industries because its critical temperature (31.1°C) allows

From: *Methods in Biotechnology, Vol. 13: Supercritical Fluid Methods and Protocols*
Edited by: J. R. Williams and A. A. Clifford © Humana Press Inc., Totowa, NJ

the processing of thermolabile organic substances, like vitamins, in an inert atmosphere, without a risk of thermal decomposition *(4)*. For liquid samples, SPE is generally preferred above liquid–liquid extraction as the isolation technique due to its speed and reduced solvent usage.

It is becoming increasingly interesting to use on-line techniques, that combine sample preparation, separation, and detection in one analytical setup. This provides a less laborious technique that is liable to automation, uses smaller amounts of sample and organic solvent, and yields enhanced analyte enrichment in a shorter time *(5–7)*. Time-consuming sample preparation steps can be eliminated resulting in faster total analysis times. Additionally, the elimination of sample handling between extraction and chromatography avoids the risk of contamination and is advantageous when labile compounds are being analyzed *(8,9)*.

For the on-line coupling of SFE and SFC, a solid phase extraction step has been employed, and is viable for aqueous samples *(10,11)*. In this way, the liquid sample is introduced in ↑ to the SFE cell filled with an adequate adsorbent, which retains the solutes of interest, while the aqueous solvent is vented with a gas purge (nitrogen). Subsequently, the analytes are extracted with supercritical carbon dioxide, and focused in a cryogenic trap, before direct injection onto the SFC column *(12)*. In addition to its fast speed, this method provides a preconcentration step for the analysis of trace levels of compounds in liquid samples. This coupled technique also allows class selective extractions based on polarity, if the extracting agent behaves as a nonpolar solvent, which in some cases also may represent an additional clean-up *(13)* to avoid interference from the sample matrix *(14)*. Therefore, the SPE-SFE of water samples with carbon dioxide will be a suitable method for the analysis of nonpolar analytes *(15)*, like fat-soluble vitamins.

In this chapter, an on-line SPE–SFE–SFC method for the analysis of fat-soluble vitamins is described. This method allows the direct introduction of large volume samples (i.e., 100–200 µL) dissolved in water or organic solvents, or in their mixtures, and may be used to analyze microdialysates *(16)*. The supercritical fluid employed was carbon dioxide, and as universal detector for the SFC, a flame ionization detector was used.

2. Materials

1. A coupled SFE–SFC system, Series 600 (Dionex, Sunnyvale, CA) equipped with on-line SFE and a Flame Ionization Detector (FID). The pump cylinder is cooled by a circulating mixture of water and ethanol at 5°C using a refrigeration bath. The SFE cell (0.3 mL, Keystone, Bellefonte, PA) is completely packed with ca. 75 mg of adsorbent. The adsorbent is a divinylbenzene:ethylvinylbenzene (55:45) polymer (Dionex, Sunnyvale, CA), with a particle size of 4.5 µm (*see* **Note 1**).

2. A CMA/200 autosampler (CMA/Microdialysis, Stockholm, Sweden), equipped with a 253 µL sample loop, delivering into the SFE cell.
3. A solvent delivery system for the automated injections, used to transfer the sample from the injection loop to the extraction solvent with the desired flow rate (*see* **Note 2**). This consists of an LC (liquid chromatography) pump and a valve coupled to the autosampler loop and operated by an LC controller (LCC-500, Pharmacia, Sweden).
4. Nitrogen (plus quality, AGA Gas AB) for drying the adsorbent and venting the water after the sample introduction and before SFE.
5. A 3-port valve (Model C3UW, Valco) connected at the end of the vent line, coupled to a linear fused silica restrictor (18 cm × 15 µm internal diameter, 144 µm o.d.), to keep the pressure in the system during SFC.
6. A cryogenic trap for concentrating the extracts on-line (*see* **Note 3**), cooled with carbon dioxide (4.8 quality, AGA Gas Gmbh, Hamburg, Germany).
7. A multiposition valve (Model CSD6UW, VICI, Valco, Houston, TX), to control the different steps of the procedure, actuated automatically (*see* **Note 4**) by an air actuator (Model A6, Valco), and a 10-port valve (Model C10W, Valco), actuated by a high temperature air actuator (Model A36-HT, Valco). These valves have 1/16 inch connections and were coupled to the autosampler and to the SFE cell, respectively, and between them, with stainless steel tubing, and to the cryogenic trap with a linear fused silica restrictor (28 cm × 15 µm internal diameter, 144 µm outer diameter).
8. An SFC open tubular column, 10 m × 50 µm internal diameter, SB-biphenyl, film thickness 0.25 µm (Dionex) coupled to a frit restrictor.
9. The assembled system containing the items described in **steps 1–8** is shown in **Fig. 1**.
10. Carbon dioxide for the extraction and chromatography, SFC-grade, was purchased from Air Liquide Gas (Malmš, Sweden) (*see* **Note 5**).
11. Organic solvents were from Merck (Darmstadt, Germany), Lichrosolv grade unless otherwise stated; ethanol (99.5%) was from Kemetyl (Haninge, Sweden), and the water was obtained through a Milli-Q water purification system (Millipore) (*see* **Note 6**).

3. Method

1. Fill the SFE cell with an adequate adsorbent previously cleaned with supercritical carbon dioxide (*see* **Note 7**). A methanol slurry of this polymer is used to pack the cell. Before the introduction of the water solution, two blank extractions are performed injecting only ethanol to check that there is no difference in the background signal obtained.
2. Inject the aqueous sample (*see* **Note 8**) automatically into the SFE cell (at 80°C) using the autosampler, and subsequently rinse the lines with ethanol. The introduced sample volume is 100 µL (*see* **Note 9**). After this, keep the cell under a nitrogen flow of ca. 60 mL/min during 15 min for solvent venting (*see* **Note 10**).

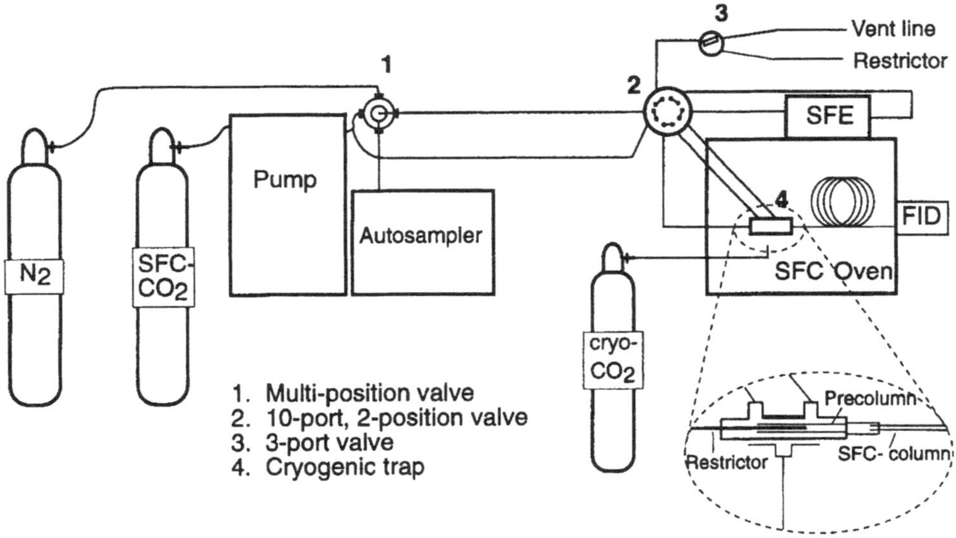

Fig. 1. Schematic diagram of the SFE-SFC system.

3. Before the SFE starts, cool the cryogenic trap in order to retain the extracted analytes and focus them until the end of the extraction. This trap is cooled with carbon dioxide (by Joule-Thompson expansion of the gas) at ca. –20°C, a temperature sufficient to lower the density of the supercritical fluid and thus reduce the solubility of the extract in the mobile phase and concentrate it in the stationary phase.
4. At the end of the solvent elimination step, switch the valves 1 and 2 to stop the nitrogen flow and to open the supercritical carbon dioxide flow. This starts the dynamic SFE. The analytes are extracted with pure carbon dioxide at 80°C, 75–400 atm at 60 atm/min, and then at 400 atm during 10 min (*see* **Note 11**), trapped and kept in the cryotrap until the start of the SFC. Hold the temperature of the SFC oven at 45°C during the extraction.
5. After the 10-min extraction, close the flow of supercritical carbon dioxide to the SFE cell with valve 1, and reduce the pressure to 100 atm. Then, close off the cooling carbon dioxide to the trap, and simultaneously raise the temperature of the oven (and the trap inside it) to 80°C. Start the SFC program, opening the flow of carbon dioxide through valve 1 and at the same time switching valve 2 to carry the supercritical carbon dioxide at 100 atm directly to the trap. In this way, the sample is quickly transferred to the column in a narrow band.
6. Carry out the SFC analysis isothermally, and start with the pressure at 100 atm, raising it by 5 atm/min to 220 atm, and then by 9 atm/min to 400 atm (*see* **Note 12**). When the program is finished, cool the SFC-oven to 45°C, switch valve 2 closing the flow of carbon dioxide, and depressurize the system from 400–75 atm at 40 atm/min). Keep the flame ionization detector at 350°C during all the procedure steps.

4. Notes

1. Critical in the procedure is the selection of the adsorbent. The selected one has a good performance not only in terms of minimum background and large breakthrough volumes for the analyzed compounds, but also a good physical stability and long durability. In this procedure, the same adsorbent could be used for more than 300 runs without any noticeable degradation. Consequently, there is no need for opening the SFE cell often to change the adsorbent, which has numerous advantages, like the saving of time and adsorbent, and minimizing the risk of contamination, leaks or irregular packing of the cell. An alternative adsorbent that showed high recoveries and no memory effects is deactivated silica (porous beads, 5-µm particles with 100-nm pores). A study of the performance of different adsorbents for this SFE–SFC coupling has been published ***(12)***.
2. The flow rate for the transfer of the sample from the sample loop to the cell with the adsorbent is very important for a high recovery. If it is too high, the analytes are not adsorbed and leave the cell through the vent line. In this application, a flow rate as high as 100 µL/min is used, although for more volatile compounds, a lower flow rate (i.e., 20 µL/min) may be needed.
3. The end of the restrictor, which is the outlet of the SFE, is located inside the cryogenic trap, inserted 3.5 cm into a deactivated fused silica precolumn. In this way, the carbon dioxide is depressurized down to atmospheric pressure in a cool environment, and consequently the extracted analytes, that are not soluble in the carbon dioxide gas (now at –20°C and 1 atm), remain in this precolumn, while the gas passes through the vent line to the atmosphere. At the same time, this uncoated precolumn (11.5 cm × 185 µm internal diameter, 340 µm outer diameter) is connected to the analytical column with a glass connector (fused silica coupler, Dionex), and focuses the analytes (i.e., concentrates them in a narrow band) at the beginning of the analytical column ***(17)***.
4. This on-line method can be performed manually switching these valves at the end of every step: solvent venting, SFE, and SFC. Nevertheless, the whole procedure is more easily carried out with the aid of a personal computer that controls automatically the switching of the valves by their respective air actuator, as shown by Ullsten and Markides ***(17)***, and which also is desirable for routine analysis.
5. During the dynamic extraction and chromatography with supercritical carbon dioxide, the fluid that exits the restrictor outlet is released to the laboratory atmosphere as a gas. The carbon dioxide is not flammable, nontoxic, environmentally nonaggressive ***(4)***, and it is not necessary to take additional precautions provided there is adequate ventilation in the laboratory.
6. Vitamins are light- and air-sensitive, and must be kept refrigerated (less than 4°C) in a colored-glass vial. The solutions were prepared and used the same day as a precaution, although no degradation is observed in solutions kept under these conditions during 2 wk.
7. The precleaning of the adsorbent is important to minimize its background signal in the SFE–SFC–FID, to take advantage of the sensitivity of the universal detector. For the recommended adsorbent, an extraction with supercritical carbon diox-

ide at 450 atm for 15 min at 50°C and another for 30 min at 100°C is enough to achieve a good and stable background signal after only three runs *(12)*. For this first extraction, an SFE-703M extractor from Dionex was used. After each chromatographic analysis of the samples, one blank run may be performed to avoid memory effects and to check the adsorbent for the next sample introduction.

8. With this set up it is possible to inject samples dissolved in organic or aqueous solvents. The injection of aqueous samples is especially important when dealing with real biological samples, for example, plasma microdialysates, and it cannot be performed with the usual injection methods in SFC. Different percentages of water in ethanol from 100–0 have been tested without any problem.
9. The sample volume may be increased to achieve a higher sensitivity. In a simple run (without making repeated injections) the only limitation is the sample loop size (in this case, 253 μL) (an accurate measure of this volume is important to get a better repeatability) and reproducible injections of 200 μL-sample may be carried out without increasing the solvent venting time (15 min). The rest of the loop volume is filled with ethanol, divided in two portions of the same volume, before and after the sample. This solvent helps to condition the adsorbent immediately before the aqueous sample reaches it, and to rinse the lines after the sample avoiding any losses of the components. For these reasons, it is convenient to employ a loop of a volume at least 20% larger than the injected sample volume.
10. The elimination of the aqueous solvent is a critical step, the drying of the adsorbent before extraction can often lead to substantial losses of volatile components *(11)*. On the other hand, the elimination of the water should be total: if some microliters of water are introduced on to the SFC column, it may cause peak distortion or even plugging of the restrictor because of the minimal solubility of water in carbon dioxide. For this reason, a nitrogen flow of 60 mL/min during venting and a solvent elimination time of 15 min were chosen, which included some additional venting time for a better performance of the method with different sample volumes. With flow rates higher than this, losses of the more volatile analytes may happen. For larger volumes of nonaqueous sample, the venting time needed may be determined by monitoring the reduction of the solvent peak in the chromatogram, but for aqueous samples, this time should be longer.
11. These conditions are enough to obtain a complete extraction of the studied lipids from this adsorbent without adding a modifier. If more polar solutes need to be extracted, a modifier such as methanol could be used.
12. It is recommended to clean the adsorbent simultaneously with the on-line SF chromatography, by using valve 2 to direct a backflush flow of supercritical carbon dioxide to the SFE cell. Consequently, the pressure of this carbon dioxide will be the same as the one of the mobile phase in the SFC. To keep the pressure in the system, valve 3 is then switched to the restrictor at the start of the SFC. This configuration is usually employed in our laboratory and is especially convenient with complex or unclean samples.

References

1. Pinkston, J. D. and Chester, T. L. (1995) Guidelines for Successful SFC/MS. *Anal. Chem.* **67,** 650A-656A.
2. Furr, H. C., Barua, A. B., and Olson, J. A. (1992) Retinoids and carotenoids, in *Modern Chromatographic Analysis of Vitamins* (De Leenheer, A. P., Lambert, W. E., and Nelis, H. J., eds.) Marcel Dekker, New York, pp. 38–39.
3. Greibrokk, T. (1995) Applications of SFE in multidimensional systems. *J. Chromatogr. A* **703,** 523–536.
4. Luque de Castro, M. D., Valcárcel, M., and Tena, M. T. (1992) *Analytical Supercritical Fluid Extraction.* Springer-Verlag, Heidelberg, pp. 62–65.
5. Lee, M. L. and Markides, K. E. (1990) *Analytical Supercritical Fluid Chromatography and Extraction.* Chromatography Conferences, Provo, Utah.
6. Louter, A. J. H., Ramalho, S., Vreuls, R. J. J., Jahr, D., and Brinkman, U. A. Th. (1996) An improved approach for on-line solid-phase extraction–gas chromatography. *J. Microcol. Sep.* **8,** 469–477.
7. Riekkola, M.-L., Manninen, P., and Hartonen, K. (1992) SFE, SFE/GC and SFE/SFC: instrumentation and applications, in *Hyphenated Techniques in Supercritical Chromatography and Extraction* (Jinno, K., ed.), Chromatography Library, Vol. 53. Elsevier Science, Amsterdam, pp. 275–304.
8. Hawthorne, S. B. (1990) Analytical-scale SFE. *Anal. Chem.* **62,** 633A-642A.
9. Chester, T. L., Pinkston, J. D., and Raynie, D. E. (1996) Supercritical fluid chromatography and extraction. *Anal. Chem.* **68,** 487R-514R.
10. Koski, I. J., Jansson, B. A., Markides, K. E., and Lee, M. L. (1991) Analysis of prostaglandins in aqueous solutions by supercritical fluid extraction and chromatography. *J. Pharm. Biomed. Anal.* **9,** 281–290.
11. Reighard, T. S. and Olesik, S. V. (1996) Bridging the gap between supercritical fluid extraction and liquid extraction techniques: alternative approaches to the extraction of solid and liquid environmental matrices. *Crit. Rev. Anal. Chem.* **26,** 61–99.
12. Petersson, U. and Markides, K. E. (1996) Stability and purity of low-polarity adsorbents for coupled supercritical fluid extraction-supercritical fluid chromatography-flame ionisation detection. *J. Chromatogr. A* **734,** 311–318.
13. Sandra, P., Medvedovici, A., Kot, A., Vilas Boas, L., and David, F. (1996) SPE-SFC-DAD: a new hyphenated system for monitoring organic micropollutants in aqueous samples. *LC-GC Int.* **9,** 540–554.
14. Pocurull, E., Marcé, R. M., Borrull, F., Bernal, J. L., Toribio, L., and Serna, M. L. (1996) On-line solid-phase extraction coupled to supercritical fluid chromatography to determine phenol and nitrophenols in water. *J. Chromatogr. A* **755,** 67–74.
15. Janda, V., Mikesová, M., and Vejrosta, J. (1996) Direct supercritical fluid extraction of water-based matrices. *J. Chromatogr. A* **733,** 35–40.
16. Señoráns, F. J., Petersson, U., and Markides, K. E. (1997) Microdialysis/SFE/SFC/FID of antioxidants and related compounds in water, in *Proceedings of the Nineteenth International Symposium on Capillary Chromatography and Electrophoresis, May 18–22, 1997,* Wintergreen, VA pp. 434–435.
17. Ullsten, U. and Markides, K. E. (1994) Automated on-line solid phase adsorption/supercritical fluid extraction/supercritical fluid chromatography of analytes from polar solvents. *J. Microcol. Sep.* **6,** 385–393.

17

Determination of Artemisinin in *Artemisia annua* L. by Off-Line Supercritical Fluid Extraction and Supercritical Fluid Chromatography Coupled to an Evaporative Light-Scattering Detector

Marcel Kohler, Werner Haerdi, Philippe Christen, and Jean-Luc Veuthey

1. Introduction

Malaria is a major disease in many countries and, according to an estimation by the World Health Organization (WHO), approx 300–500 million people contract malaria yearly and almost 2 million die annually *(1)*. Controlling malaria is now becoming very problematic because of the developing resistance of *Plasmodium falciparum* to chloroquine, mefloquine, and other commonly used antimalarial drugs.

Artemisinin is a promising drug against chloroquine-resistant strains of *Plasmodium* and in the treatment of cerebral malaria *(2–4)*. This compound is an endoperoxide sesquiterpene lactone found in the aerial parts of the plant *Artemisia annua* L.(Asteraceae), a plant that has been used for many centuries in traditional Chinese medicine for the treatment of fever and malaria. Although the total synthesis of artemisinin has been achieved *(5)*, it is not as competitive in price as the natural product. The concentration of artemisinin, obtained from cultivated *A. annua,* varies in the range of 0.01% to around 1% of the plant's dry weight *(3,4,6)* and levels depend on many factors, such as the plant's origin, its stage of development and the cultivation conditions. Hence it is necessary to use analytical methods that can detect artemisinin and its major bioprecursor, artemisinic acid (**Fig. 1**), in the plant.

A number of analytical methods exist for determining artemisinin and its derivatives, such as high-performance liquid chromatography (HPLC) coupled

From: *Methods in Biotechnology, Vol. 13: Supercritical Fluid Methods and Protocols*
Edited by: J. R. Williams and A. A. Clifford © Humana Press Inc., Totowa, NJ

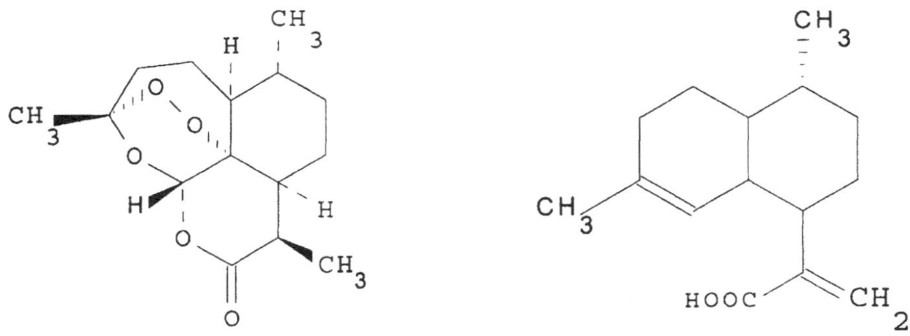

ARTEMISININ ARTEMISINIC ACID

Fig. 1. Structure of artemisinin and artemisinic acid.

with ultraviolet detection (UV) *(7)*, electrochemical detection (EC) *(8)*, mass spectrometry (MS) *(9)*, thin-layer chromatography (TLC) *(10)*, gas chromatography (GC) *(11)*, GC-MS *(12)*, and enzyme-immunoassay *(13)*. Very few of these methods allow a simultaneous and direct determination of artemisinin, artemisinic acid, and other derivatives. Indeed, artemisinin is a thermolabile compound that cannot be determined without degradation by GC. Therefore, GC and GC-MS analyses measure artemisinin indirectly by detecting its degradation products. Artemisinin is UV-transparent and requires a derivatization before HPLC-UV analysis. However, HPLC-EC measures artemisinin directly, as well as the derivatives which possess an endoperoxide bridge such as artemisitene, but artemisinic acid cannot be determined by this method. Finally, even if HPLC-MS can detect artemisinin and its derivatives directly and simultaneously, this technique is not currently used in many analytical laboratories and remains costly.

Therefore, it becomes inevitable to look for an alternative method that can determine simultaneously artemisinin and artemisinic acid in crude *A. annua* extracts. Artemisinin is an excellent candidate for supercritical fluid chromatography or SFC (*see* Chapter 1), a technique that emerged in the 1980s as a very powerful method, complementary to GC and HPLC. Because SFC allows to work at low temperature, no degradation of artemisinin is observed and fast analyses can be carried out due to the large diffusion coefficients of analytes in supercritical fluids *(14)*. Furthermore, universal detectors used currently in chromatography, such as the evaporative light-scattering detector (ELSD), can be coupled to SFC *(15)*.

The phenomenon of light scattering has been used for many years in a large variety of measurements and has been applied more recently to a chromato-

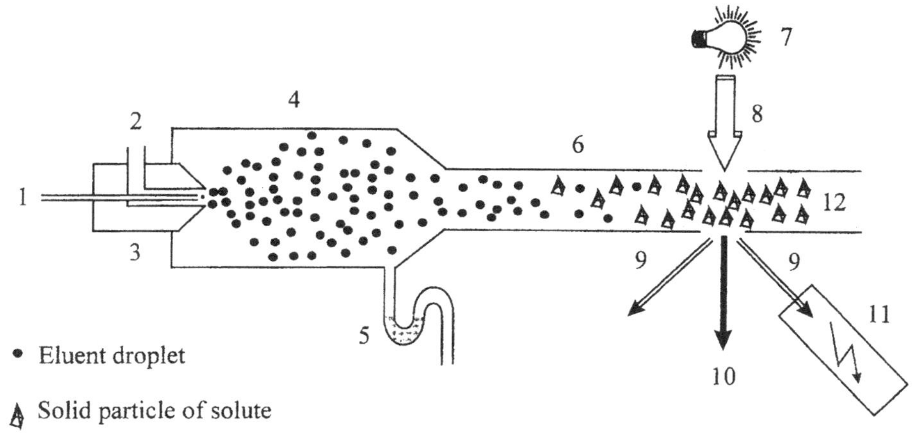

Fig. 2. Schematic representation of an ELSD: *1,* HPLC effluent; *2,* nebulizing gas; *3,* concentric nebulizer; *4,* nebulizing chamber; *5,* liquid waste (settled droplets); *6,* heated drift tube; *7,* light source; *8,* light beam; *9,* diffracted light; *10,* transmitted light; *11,* photomultiplier; *12,* gas exhaust.

graphic detector. Schematically (**Fig. 2**), the effluent from a chromatographic column enters a nebulizer where it is converted to an aerosol with the aid of a carrier gas. The fine droplets are then carried into a heated drift tube where the solvent is evaporated to form small particles of pure solute. At the end of the drift tube, a light beam is scattered by the particles present in the gas flow and the scattered light is detected by a photomultiplier. The measured light is proportional to the amount of sample and is not dependent on a specific functional group or chromophore. Contrary to the refractive index detector, ELSD is not sensitive to temperature fluctuation and can be used with gradient elution without significant baseline drift. However, this detector is limited by the complete volatilization of all mobile phase components. The ELSD allows direct detection of all nonvolatile compounds, regardless of their chemical structure, and is therefore a valuable tool in the determination of compounds without chromophores ***(16)***.

Whatever the analytical method used, an extraction procedure of the plant material is required. Liquid solvent extraction with toluene, hexane, or petroleum ether is the most currently applied technique, with extraction times that can vary from a few minutes to several hours. However, these procedures use a large amount of potentially hazardous solvents, which have to be eliminated before analysis. Therefore, in view of its properties already described in the literature ***(17–19)***, supercritical fluid extraction (SFE) with carbon dioxide is an interesting alternative to conventional liquid solvent extraction methods,

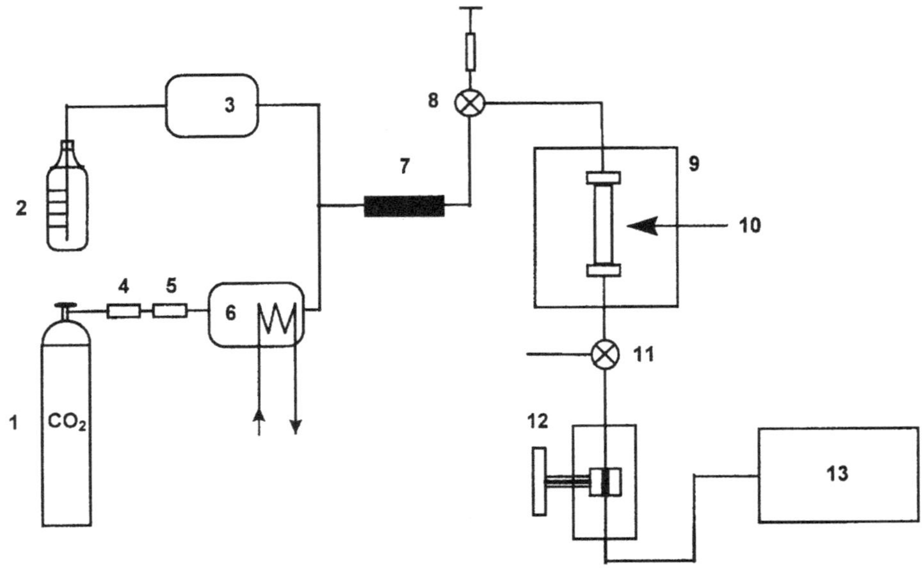

Fig. 3. Schematic diagram of the SFC-ELSD system. *1*, CO_2 cylinder with eductor tube; *2*, polar modifier; *3*, pump; *4*, charcoal-packed column; *5*, molecular sieve-packed column; *6*, pump with cooling jacket; *7*, glass beds column; *8*, switching valve; *9*, oven; *10*, chromatographic column; *11*, purge valve; *12*, pinched restrictor; *13*, ELSD.

especially in the case of plant material ***(20–22)***. Sesquiterpene lactones, such as artemisinin, are slightly polar compounds, which can be extracted by supercritical fluids. Recently, we showed ***(23)*** that artemisinin could be extracted from *Artemisia annua* with carbon dioxide and a small addition of methanol or ethanol was sufficient to achieve a rapid and quantitative extraction, whatever the pressure and the temperature used.

Thus, in this chapter, we present an SFC-ELSD method that determines artemisinin and artemisinic acid without derivatization and without decomposition in plant extracts. These latter were obtained by supercritical fluid extraction.

2. Materials

1. Carbon dioxide, 99.99% pure, or CO_2 for SFC (Polygaz, Geneva, Switzerland) contained in a cylinder with an eductor tube. Analyses are performed with a Varian 2510 HPLC pump (Varian, Palo Alto, CA) equipped with a cooling jacket for CO_2, and the polar modifier is added through a T junction with a Knauer HPLC pump 64 (Knauer, Berlin, Germany). Analyses are performed on a packed column (*see* **Subheading 3., step 1**). The column is coupled to the Sedex 55 ELSD (S.E.D.E.R.E, Alfortville, France) through a homemade restrictor (*see* **Note 1**) as shown in **Fig. 3**.

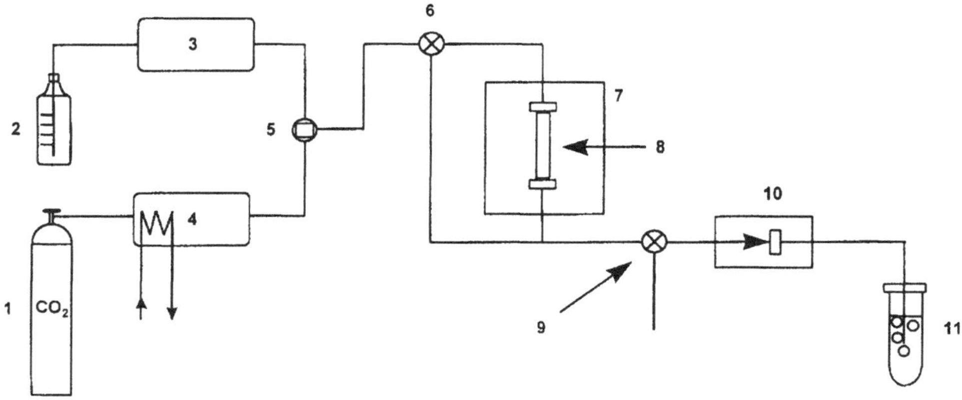

Fig. 4. Instrumental set-up for supercritical fluid extraction. *1*, CO_2 cylinder with eductor tube; *2*, polar modifier; *3*, pump; *4*, pump with cooling jacket; *5*, dynamic mixer; *6*, valve; *7*, oven; *8*, extraction cell; *9*, purge valve; *10*, variable restrictor; *11*, collection vial.

2. HPLC-grade (*see* **Note 2**) methanol, acetonitrile, and ethanol are purchased from Maechler AG (Basel, Switzerland). Crystalline artemisinin is obtained from Sigma SA (Sigma, St. Louis). Artemisinic acid is kindly provided by Dr. N. Acton (Walter Reed Army Institute of Research, Washington D.C.). Stock solutions of artemisinin (10 mg/mL) and artemisinic acid (5 mg/mL) are made in acetonitrile and are stored at 4°C for up to 3 mo. Standard solutions containing artemisinin and artemisinic acid are prepared daily by diluting the stock solution with acetonitrile.
3. Supercritical fluid extraction (SFE) of plant material is conducted in a 1 mL (14 mm × 10 mm ID) Jasco extraction cell (Tokyo, Japan). The temperature is regulated by a column oven (Jasco CO-965). The CO_2 and ethanol (as modifier) are pumped by two HPLC pumps operated in constant flow mode (Jasco PU-980) as shown in **Fig. 4**. Authentic plant material is kindly provided by Dr. N. Delabays (Mediplant, Conthey, Switzerland).

3. Method

1. For SFC, a charcoal-packed column and a molecular sieve-packed column are incorporated between the cylinder and the pump to prevent possible contamination by hydrocarbons present in CO_2. The supercritical fluid (CO_2 and methanol as modifier) is homogenized by passing through a preliminary column (150 mm × 4 mm ID) filled with glass beds (1 mm diameter). The sample (20 µL) is injected into the chromatographic column (a Nucleosil 100-5 NH_2, 200 mm × 4 mm ID by Macherey-Nagel, Oensingen, Switzerland), which is coupled to the ELSD through a pinched peek restrictor (homemade) heated at 80°C to avoid dry ice formation. The analysis is carried out at a temperature of 40°C, using a polar modifier (methanol) gradient. Initially, 1% methanol is added to the CO_2. After 3 min, the methanol

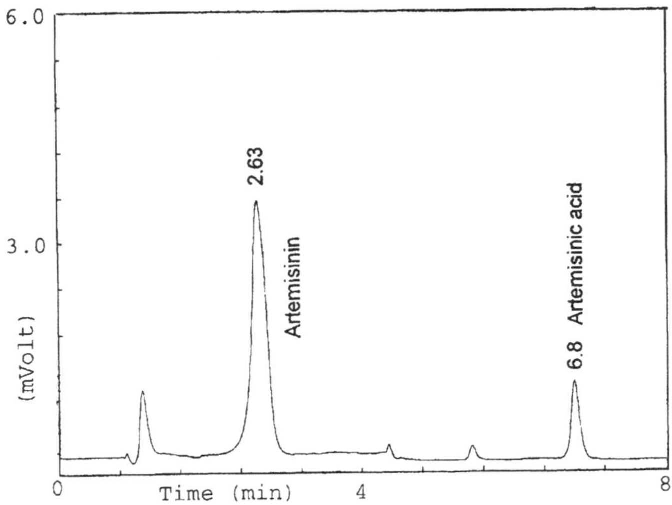

Fig. 5. SFC-ELSD chromatogram of a standard solution of artemisinin and artemisinic acid using a polar modifier gradient. Initial 1%, after 3 min the methanol percentage was enhanced to 10% in 12 s and held for 5 min. The flow was set at 4 mL/min and the pressure was set at 170 bar.

percentage is enhanced to 10% in 0.2 min and held for 5 min. The flow rate is set at 4 mL/min and the pressure is set at 170 bar. The conditions of the ELSD are: air pressure 0.5 bar (6 L/min), temperatures of the nebulization chamber and the heated drift tube chamber are set at 40°C. Integration is done by a Hewlett-Packard 3396 series II integrator.

2. For validation of the method, a calibration curve is produced for concentrations between 0.1 to 1.0 mg/mL ($n = 5$). Because the response of the ELSD is related to the concentration of the analyte through an exponential relation (*see* **Note 3**), logarithms of peak areas of artemisinin and artemisinic acid are reported as a function of their concentrations (in logarithms). For both compounds, the linearity is verified (correlation coefficients are greater than 0.99) and repeatabilities ($n = 6$), expressed by the relative standard deviations, are inferior to 8%. In the optimized analytical conditions, artemisinin and artemisinic acid are separated in less than 8 min with retention factors of 1.4 and 5.1, respectively (**Fig. 5**).
3. For SFE, the air-dried plant material is thoroughly ground (470 μm) in a domestic mixer. A sample of this material (100 mg) is introduced into the extraction cell. The temperature of this cell is set at 50°C. The supercritical fluid (CO_2 and 3% ethanol) is pumped at a flow rate of 2 mL/min (expressed as the sum of liquid CO_2 and modifier). The pressure in the system is regulated at 150 bar through a variable restrictor (Jasco 880-01 Back Pressure Regulator). This latter is heated at 50°C to avoid dry ice formation and the sample is collected by bubbling into 5 mL of ethanol contained in a 15 mL conical centrifuge tube maintained at 25°C

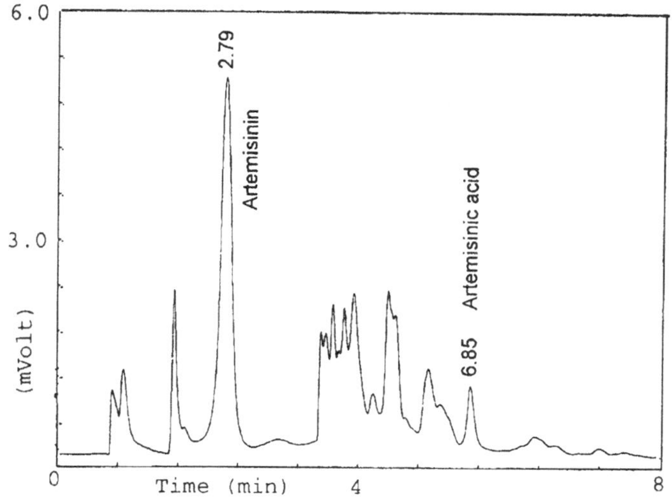

Fig. 6. SFC-ELSD chromatogram of an *Artemisia annua* L. extract using the same conditions as described in **Fig. 5**.

(*see* **Note 4**). The extract is evaporated to dryness under a nitrogen flow at 40°C and the dry residue is redissolved in 500 μL of acetonitrile. This solution is filtered through a 0.22 μm membrane and is ready to be injected in duplicate (**Fig. 6**). For quantitative determination, three standard solutions containing artemisinin (200, 400, and 600 ppm) and artemisinic acid (50, 100, and 200 ppm) are injected at the beginning and at the end of a sequence to plot a calibration curve. Each sequence consists of four plant extracts.

4. Notes

1. In order to have a better control on the pressure and the flow rate and to minimize dead volumes between the column and the detector, a 100 μm ID × 10 cm length of PEEK tubing (Upchurch Scientific, Oak Harbor, WA) is used as a restrictor. The extremity of the tubing is inserted between two stainless steel disks of 1 cm diameter which can be heated, tightened by means of a micrometric screw and placed directly in the nebulization chamber of the ELSD. The advantage of this restrictor is that PEEK material regains its original form even after being strongly pressed, therefore, only one tubing can be used to set the chromatographic pressure. Furthermore, this system is cost-effective with regard to other restrictors commercially available.
2. In order to obtain a low noise background with the ELSD, mobile phases have to be constituted of high-grade solvents (without dry residues) and of volatile buffer components.
3. Due to the response of the interactions involved, the response of the ELSD cannot be related to the mass of the analyte by a linear equation. In fact, the response is rather exponential:

$$\text{Signal} = C_1 \times m^{C_2}$$

where m represents the mass of the injected analyte, C_1 and C_2 are two constants determined principally by the nature of the mobile phase. C_2 generally varies between 1 and 2 depending on the conception of the apparatus. Therefore, calibration curves are plotted on a double logarithmic scale. According to our own experience, the linearity is not applicable over two orders of magnitude.

4. The extracted compounds are not lost by aerosol formation.

References

1. World Health Organization (1996) *World Malaria Situation in 1993*, World Health Organization, Geneva.
2. Klayman, D. L. (1985) Qinghaosu (artemisin): an antimalarial drug from China. *Science* **28,** 1049–1055.
3. Woerdenbag, H. J., Lugt, C. B., and Pras, N. (1990) Artemisia annua L.: a source of novel antimalarial drugs. *Pharm. Weekbl. Sci. Ed.* **12,** 169–181.
4. Hien, T. T. and White, N. J. (1993) Qinghaosu. *Lancet* **341,** 603–608.
5. Schmid, G. (1983) Total synthesis of Qinghaosu. *J. Am. Chem. Soc.* **105,** 624–625.
6. Woerdenbag, H. J., Pras, N., Chan, N. G., Bang, B. T., Bos, R., Van Uden, W., Van, Y. P., Boi, N. V., Batterman, S., and Lugt, C. B. (1994) Artemisinin, related sesquiterpenes, and essential oil in *Artemisia annua* during a vegetation period in Vietnam. *Planta Med.* **60,** 272–275.
7. Shisan, Z. and Mei-Yi, Z. (1986) Application of precolumn reaction to high performance liquid chromatography of Qinghaosu in animal plasma. *Anal. Chem.* **58,** 289–292.
8. Acton, N., Klayman, D. L., and Rollman, I. J. (1985) Reductive electrochemical HPLC assay for artemisinin (Qinghaosu). *Planta Med.* **51,** 445–446.
9. Leskovac, V., and Theoharides, A. D. (1991) Hepatic metabolism of artemisinin drugs. I. Drug metabolism in rat liver microsomes. *Comp. Biochem. Physiol.* **99C,** 383–396.
10. Pras, N., Visser, J. F., Batterman, S., Woerdenbag, H. J., Malingré, T. M., and Lugt, C. B. (1991) Laboratory selection of Artemisia annua L. for high yielding types. *Phytochem. Anal.* **2,** 80–83.
11. Sipahimalani, A. T., Fulzele, D., and Heble, M. R. (1991) Rapid method for the detection and determination of artemisinin by gas chromatography. *J. Chromatogr.* **538,** 452–455.
12. Woerdenbag, H. J., Pras, N., Bos, R., Visser, J. F., Hendriks, H., and Malingré, T. M. (1991) Analysis of Artemisinin and related sesquiterpenoids from *Artemisia annua* L. by combined gas chromatography/mass spectrometry. *Phytochem. Anal.* **2,** 215–219.
13. Jaziri, M., Diallo, B., Vanhaellen, M., Homès, J., Yoshimatsu, K., and Shimomura, K. (1993) Immunodetection of artemisinin in *Artemisia annua* cultivated in hydroponic conditions. *Phytochemistry* **33,** 821–826.
14. Chester, T. L., Pinkeston, J. D., and Raynie, D. E. (1994) Supercritical fluid chromatography and extraction. *Anal. Chem.* **66,** 106R-130R.

15. Thompson, J., Strode, B., and Taylor, L. T. (1996) Evaporative light scattering detector for supercritical fluid chromatography. *J. Chrom. Sci.* **34,** 261–271.
16. Kohler, M., Haerdi, W., Christen, P., and Veuthey, J.-L. (1997) The evaporative light scattering detector: some applications in pharmaceutical analysis. *Trends Anal. Chem.* **16,** 475–484.
17. King, M. B. and Bott, T. R., ed. (1995) *Extraction of Natural Products Using Near-Critical Solvents.* Blackie Academic and Professional, London.
18. King, J. and France, J. E. (1992) Basic principles of analytical supercritical fluid extraction, in *Analysis With Supercritical Fluids: Extraction and Chromatography* (Wenclawiak, B., ed.), Springer Laboratory, Berlin, pp. 32–60.
19. Hawthorne, S. B. (1993) Methodology for off-line supercritical fluid extraction, in *Supercritical Fluid Extraction and Its Use in Chromatographic Sample Preparation* (Westwood, S. A., ed.), Blackie Academic and Professional, London.
20. Castioni, P., Christen, P., and Veuthey, J.-L. (1995) L'Extraction en phase supercritique des substances d'origine végétale. *Analusis* **23,** 95–106.
21. Bevan, C. D. and Marshall, P. S. (1994) The use of supercritical fluids in the isolation of natural products. *Nat. Prod. Rep.* **11,** 451–466.
22. Modey, W. K., Mulholland, D. A., and Raynor, M. W. (1996) Analytical supercritical fluid extraction of natural products. *Phytochem. Anal.* **7,** 1–15.
23. Kohler, M., Haerdi, W., Christen, P., and Veuthey, J.-L. (1997) Supercritical fluid extraction and chromatography of artemisinin and artemisinic acid: an improved method for the analysis of *Artemisia annua* samples. *Phytochem. Anal.* **8,** 223–227.

18

Analysis of *Cannabis* by Supercritical Fluid Chromatography with Ultraviolet Detection

Michael D. Cole

1. Introduction

Cannabis sativa L. and its products comprise a significant and important part of the forensic drug laboratory's case load. Two principle types of analyses are required for the analysis of *Cannabis*, namely, identification of the material, since it is a substance controlled in the United Kingdom under the Misuse of Drugs Act, 1971 and its amendments, and second, the comparison of two or more samples of *Cannabis* to determine if they once formed a larger sample *(1,2)*. Such analyses are generally carried out using combinations of presumptive tests, thin-layer chromatography (TLC), high-performance liquid chromatography (HPLC), gas chromatography (GC) and gas chromatography–mass spectroscopy (GC–MS) *(2–4)*.

Whilst TLC is rapid and inexpensive, it is neither definitive nor accurately quantitative. HPLC offers greater resolution than TLC, but suffers from long analysis times, short analytical column life and is not definitive. GC and GC-MS offer the greatest resolution of the components of the samples, but require derivatization of the samples before analysis for complete comparison of the samples (because of the thermal lability of some of the components of the mixture) and, hence, suffer from all of the problems concomitant with such procedures.

Supercritical fluid chromatography (SFC) has been employed for a number of different analyses of drugs of abuse, including barbiturates *(5)*, benzodiazepines *(6)*, opiates *(7)*, cocaine *(8)*, and cannabinoid metabolites *(9)*. SFC offers greater resolution than HPLC, but without the need to derivatize the samples. Coupled to atmospheric pressure chemical ionization mass spectroscopy (APCI-MS), the technique offers definitive identification of the analytes *(10)*.

From: *Methods in Biotechnology, Vol. 13: Supercritical Fluid Methods and Protocols*
Edited by: J. R. Williams and A. A. Clifford © Humana Press Inc., Totowa, NJ

In this chapter, a simple method for the SFC-UV analysis of *Cannabis* is described providing an alternative means for the identification and comparison of *Cannabis* samples. When the method described is hyphenated to a mass spectrometer with an APCI source, an attractive definitive technique for the identification of *Cannabis* is provided.

2. Materials

1. Authenticated cannabinoid standards (Δ^9-tetrahydrocannabinol, Δ^8-tetrahydrocannabinol, cannabinol, cannabidiol) at 1 mg/mL in ethanol (*see* **Note 1**).
2. Analytical reagent grade ethanol.
3. Pestle and mortar.
4. A supply of 6-dram vials.
5. A microfuge and tubes.
6. An electronic balance.
7. A supercritical fluid chromatograph with the ability to deliver CO_2 and methanol, fitted with a 5-µL injection loop and interfaced to an ultraviolet detector.
8. A cyanopropyl silica column (25 cm × 4.6 mm internal diameter, packed with 5-µm spherical particles).

3. Method

1. Prepare the standard solutions of the four compounds, listed in the previous section, at a concentration of 1 mg/mL in ethanol, since this is the solvent in which the cannabinoids are most stable. The solutions should be freshly prepared and stored at 4°C in the dark. This minimizes the risk of decomposition of the Δ^9-tetrahydrocannabinol into cannabinol.
2. Grind the *Cannabis* products to be analyzed (herbal material or resin) to a fine powder in the pestle and mortar. Following this, the powder should be triturated in ethanol at a concentration of 10 mg/mL, and extracted for 10 min at room temperature. The extracted materials should be transferred to microfuge tubes and centrifuged at 4000g for 5 min. This removes the solid material from the extract. The supernatant should be carefully removed for analysis, and great care should be taken to ensure that the plant material pellet is not disturbed. The samples should be stored at 4°C in the dark before analysis, to minimize the risk of chemical decomposition.
3. Also prepare a solvent control (blank) to demonstrate that any extracted compounds arise from the extraction of *Cannabis* and not from the plastic of the microfuge tubes.
4. Prepare the SFC system and allow it to equilibrate for 30 min at the start of each day. The analysis is performed by using a mobile phase flow rate of 2 mL/min, at a pressure of 3000 psi, using 2% methanol by volume in CO_2 initially rising linearly to 7% at 15 min. The eluate should be monitored with the UV detector at 210 nm.
5. Confirm the correct functioning of the instrument at the start of each day by analyzing a freshly prepared standard drug solution. The elution order in the

system described is cannabidiol, Δ^8-tetrahydrocannabinol, Δ^9-tetrahydrocannabinol, and cannabinol. The absolute retention times will vary with system and operator, but exemplar values will be in the region of 4.2, 4.7. 5.2, and 7.0 min, respectively (*see* **Note 2**).

6. Confirm the cleanliness of the system by the analysis of an ethanol blank. A straight baseline demonstrates that there has been no carry-over between analyses.
7. The samples should be analyzed by using the same procedure. Between each sample analysis a solvent blank, treated in a microfuge tube, should be analyzed, to demonstrate that carry-over has not occurred between samples.
8. Due to the complex nature of plant and natural products, an analysis of the standard solution should be made between every fourth or fifth sample to demonstrate that the column is still functioning correctly (*see* **Notes 3–5**).
9. Identify the compounds on the basis of retention time data. Comparison can also be made between the chromatograms obtained from each sample to determine whether the drug samples are related to each other.

4. Notes

1. If the standard solution starts to become brown and discolored, this suggests that the standard compounds are decomposing. This solution should be replaced before proceeding.
2. If split peaks or double peaks are observed during the analysis, it has been our experience that the problem can be overcome by diluting the sample. It is hypothesized that some of the cannabinoid sample precipitates, becomes trapped at the top of the column and then redissolves in the supercritical CO_2. The result of this is that a double peak for the analytes is observed.
3. Since the *Cannabis* products can include complex mixtures of lipids and phenolics, regular washing of the column at the end of each day is recommended. In our laboratory washing the column with supercritical CO_2 modified with 20% methanol has been found to be effective.
4. When producing calibration curves for the quantitative determination of the cannabinoids in the sample, the solutions should be analyzed from the lowest concentration to the highest concentration in ascending order, with each sample separated by a blank. This is necessary in forensic science to prevent saturation of the analytical column and to demonstrate that there has been no carryover between samples.
5. Due to the complex nature of *Cannabis* products, it is possible that residues from the extracts can accumulate on moving parts and the pressure regulators of the SFC system. It is our experience that regular instrument maintenance and cleaning of these components with absolute ethanol alleviates this problem.

References

1. Gough, T. A. (1991) *The Analysis of Drugs of Abuse.* Wiley, Chichester, UK.
2. Anon (1992) *Recommended Methods for the Testing of* Cannabis. United Nations Drug Control Programme, Vienna and New York.

3. Lehmann, T. and Brenneisen, R. (1995) High performance liquid chromatographic profiling of *Cannabis* products. *J. Liq. Chromatogr.* **187,** 689–700.
4. Huizer, H. (1991) The use of gas chromatography for the detection and quantification of abused drugs, in *The Analysis of Drugs of Abuse* (Gough, T. A., ed.), Wiley, Chichester, UK.
5. Smith, R. M. and Sanagi, M. M. (1989) Supercritical fluid chromatography of barbiturates. *J. Chromatogr.* **481,** 63–69.
6. Smith, R. M. and Sanagi. M. M. (1989) Packed column supercritical fluid chromatography of benzodiazepines. *J. Chromatogr.* **483,** 51-61.
7. Janicot, J. L., Caude, M., and Rosset, R. (1998) Separation of opium alkaloids by carbon dioxide subcritical and supercriticial fluid chromatography with packed columns-application to the quantitative analysis of poppy straw extracts. *J. Chromatogr.* **437,** 351–364.
8. Mackay, G. A. and Reed, G. D. (1991) The application of capillary SFC, packed column SFC and capillary SFC-MS in the analysis of controlled drugs. *J. High Res. Chromatogr.* **14,** 537–541.
9. Later, D. W., Richter, B. E., Knowles, D. E., and Anderson, M. R. (1986) Analysis of various classes of drugs by capillary supercritical fluid chromatography. *J. Chromatogr. Sci.* **24,** 249–253.
10. Backstrom, B., Cole, M. D., Carrott, M. J., Jones, D. C., Davidson. G., and Coleman, K. (1997) A preliminary study of the analysis of *Cannabis* by supercritical fluid chromatography with atmospheric pressure chemical ionisation mass spectroscopic detection. *Science Justice* **37,** 91–97.

19

Direct Chiral Resolution of Optical Isomers of Diltiazem Hydrochloride by Packed Column Supercritical Fluid Chromatography

Koji Yaku, Keiichi Aoe, Noriyuki Nishimura, Tadashi Sato, and Fujio Morishita

1. Introduction

Packed column subcritical and/or supercritical fluid chromatography (p-sub- or pSFC) has been used as a powerful chiral separation technique, whereby a mobile phase produces low viscosity, a high diffusion coefficient, and a solvating power. P-sub- or p-SFC tends to obtain higher column efficiency than normal-phase high-performance liquid chromatography (HPLC). Chiral separations using p-sub- or p-SFC, as well as HPLC, have been reported by many researchers, who frequently use columns containing derivatized cellulose packing *(1–10)*.

Diltiazem hydrochloride, (2*S*,3*S*)-3-acetoxy-2,3-dihydro-2-(4-methoxyphenyl)-5-(2dimethylaminoethyl)-1,5-benzothiazepine-4(5*H*)-one monohydrochloride (shown in **Fig. 2**), is a benzothiazepine-type Ca-antagonist developed originally by Tanabe Seiyaku Co. It has been widely used worldwide for the treatment of angina pectoris, variant angina, and essential hypertension, which are attributable to the Ca-antagonistic action. Diltiazem hydrochloride has asymmetric carbons at positions 2 and 3. There are two isomers, *cis* and *trans*, depending on the relative positions of the substituents. Each isomer also has *d*- and *l*- optical isomers. Diltiazem hydrochloride is a *d*-*cis*-(2*S*,3*S*) isomer. It is known that, in general, the determination of the optical impurity in the drug is very important from the efficacy and safety point of view. The methods of separation for optical isomers of diltiazem hydrochloride by reversed- and normal-phase HPLC have already been reported *(11–15)*. In p-SFC, the chiral resolution of four optical isomers of diltiazem hydrochloride has been optimized

From: *Methods in Biotechnology, Vol. 13: Supercritical Fluid Methods and Protocols*
Edited by: J. R. Williams and A. A. Clifford © Humana Press Inc., Totowa, NJ

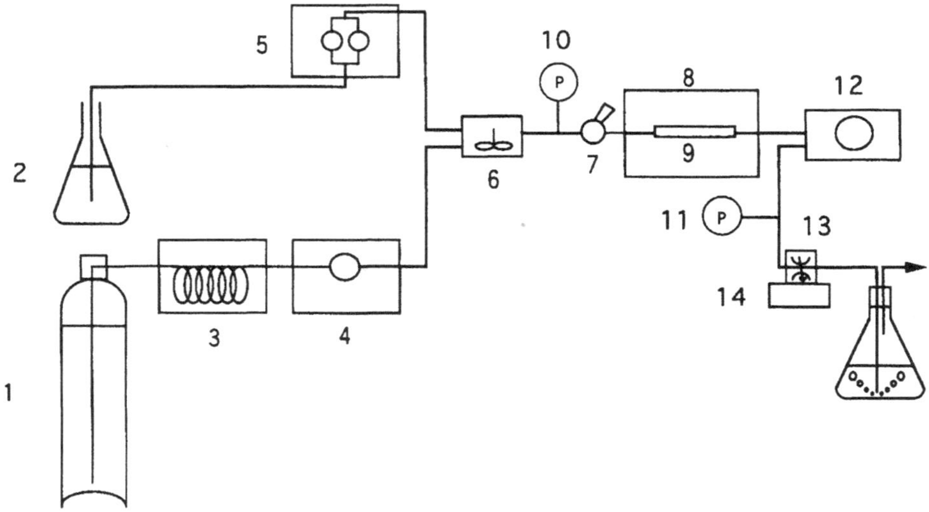

Fig. 1. Diagram of the packed-column supercritical fluid chromatograph. *1,* carbon dioxide cylinder; *2,* reservoir for the modifier; *3,* cooler; *4,* pump for delivering carbon dioxide *5,* pump for delivering the modifier; *6,* dynamic mixer for mixing the modifier and carbon dioxide; *7,* injection valve fitted with a 5 µL sample loop; *10* and *11,* pressure monitors; *12,* UV detector; *13,* back-pressure regulator; and *14,* dry thermo unit for heating the back-pressure regulator.

based on the evaluation of the effects of columns, modifiers and additives, pressure, and temperature ***(16)***. The optical isomers were separated with baseline resolution on a Chiralcel OD column within 8 min, indicating high column efficiency.

The protocols of the instrumentation of p-SFC (modified from a commercial HPLC apparatus) and the optimum chiral resolution methods are presented in detail in this chapter. In addition, the determination of three optical impurities in diltiazem hydrochloride and comparison with HPLC separation are also described briefly (*see* **Notes 1** and **2**).

2. Materials

1. A high-performance liquid chromatograph modified for p-SFC operation ***(17)*** as shown in **Fig. 1**. This is comprising
   a. a carbon dioxide cylinder with a dip tube for delivering liquid (*see* **step 4**)
   b. a reservoir for the modifier
   c. a 1.6-mm outer diameter coiled stainless steel tube situated in a cooling bath acting as a heat exchanger
   d. a pump, such as a single-plunger reciprocating pump (e.g., model LC-6A, Shimadzu, Kyoto, Japan), fitted with a jacket around the pump head for circulating coolant, for delivering carbon dioxide

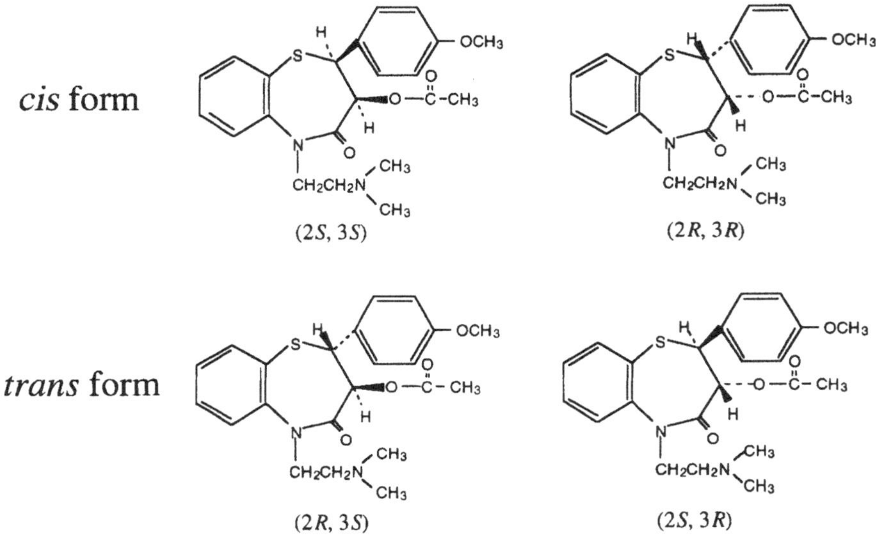

Fig. 2. Chemical structures of diltiazem optical isomers.

   e. a pump for delivering the modifier, such as a double-plunger reciprocating pump (e.g., model LC-9A, Shimadzu)
   f. a dynamic mixer for mixing the modifier and carbon dioxide, (e.g., model MX-8010, Tosoh, Osaka, Japan) with a chamber volume of 1.9 mL and a maximum working pressure of 400 bar
   g. an injection valve fitted with a 5-µL sample loop (e.g., model 8125, Rheodyne, Cotati, CA)
   h. an oven (e.g., model CTO-6A, Shimadzu)
   i. and j. pressure monitors (e.g., model LC-6AD, Shimadzu)
   k. a UV detector with a flow cell of volume 3 µL and a maximum working pressure of 400 bar (e.g., model SPD-6A, Shimadzu)
   l. a back-pressure regulator (e.g., model 26-1722-24-043, Tescom Instruments, Elk River, MI) (*see* **Note 3**)
   m. a dry thermo unit for heating the back-pressure regulator (e.g., TAL-1G, TAITEC, Osaka, Japan) (*see* **Note 4**).
2. A circulating cooling bath for cooling the heat exchanger and pump head (e.g., cooling pump CH-150B and pump unit P-1, TAITEC).
3. A Chiralcel OD column (250 mm × 4.6 mm internal diameter, packing particle size 10 µm, Daicel Chemicals, Tokyo, Japan) (*see* **Note 5**).
4. A cylinder of carbon dioxide of more than 99.9% purity fitted with a diptube.
5. Isopropanol and diethylamine of HPLC grade or analytical reagent grade.
6. Diltiazem hydrochloride and its three isomers, shown in **Fig. 2**, (synthesized by the Tanabe Seiyaku Co., Osaka, Japan).
7. An integrator (e.g., Chromatopac C-R5A integrator, Shimadzu), to record the results.

3. Method

1. Dissolve diltiazem hydrochloride and its three isomers in ethanol at ca. 1 mg/mL.
2. Cool the cooling bath (**Subheading 2, step 2**) and pump head of the carbon dioxide pump (**Subheading 2, step 1**) to –10°C.
3. Set the oven (column) temperature to 50°C in the oven (*see* **Note 6**).
4. Set the temperature of the back-pressure regulator to 40°C with the heating-unit (*see* **Note 4**).
5. Set the wavelength of the UV detector to 254 nm.
6. After the pump head has reached the temperature of –10°C, pump the liquid carbon dioxide with a flow-rate of 2 mL/min (*see* **Note 7**).
7. Adjust the outlet pressure to 180 bar with the back-pressure regulator.
8. Pump isopropanol containing 0.5% v/v diethylamine at a flow-rate of 0.3 mL/min (*see* **Notes 8** and **9**).
9. After the system reaches equilibrium, inject the sample for analysis (*see* **Note 10**). As shown in **Fig. 3A**, diltiazem and its three optical isomers are resolved at the baseline on a Chiralcel OD column within 8 min (*see* **Notes 11–13**).

4. Notes

1. An example of the determination of the three optical isomer impurities spiked into in diltiazem hydrochloride is shown in **Fig. 4**. The limit of detection was found to be 0.05% of impurities in diltiazem, as shown in **Fig. 4A**. Replicate separations of diltiazem containing 1% of the isomer impurities, shown in **Fig. 4B**, were found to be of good precision. Linearity was also found to be good. Analysis of bulk product drugs showed an absence of optical isomer impurities, i.e., less than 0.05%.
2. Comparison was made with HPLC separation. The separations of the four optical isomers on the Chiralcel OD and OF columns obtained by p-SFC (shown in **Fig. 3**) are compared with those obtained by HPLC (shown in **Fig. 5**). It can be seen that, the *d-trans* and *l-cis* isomers are not resolved on the Chiralcel OD column in HPLC, although they are the geometric isomers. On the Chiralcel OF column, all isomers achieved baseline separation in both modes, but the elution order of *d-trans* and *l-cis* isomers in p-SFC and HPLC are different. The plate numbers obtained in p-SFC are higher by a factor of 2–3.8 in comparison with those in HPLC; 2022-6137 in p-SFC and 539-3223 in HPLC. Thus, higher efficiencies for the chiral separation of diltiazem hydrochloride can be obtained in p-SFC than in HPLC, especially on a Chiralcel OD column, for which the most rapid separations are obtained.
3. Since the pressure is controlled by the back-pressure regulator, the inlet flow-rate can be changed independently of the pressure.
4. To prevent clogging with solid carbon dioxide, the regulator should be heated.
5. The stationary phases are cellulose derivatives coated on to a silica support, which are cellulose tris(3,5-dimethylphenylcarbamate), cellulose tris(phenylcarbamate), and cellulose tris(4-chlorophenylcarbamate) for the Chiralcel OD, OC, and OF columns, respectively.

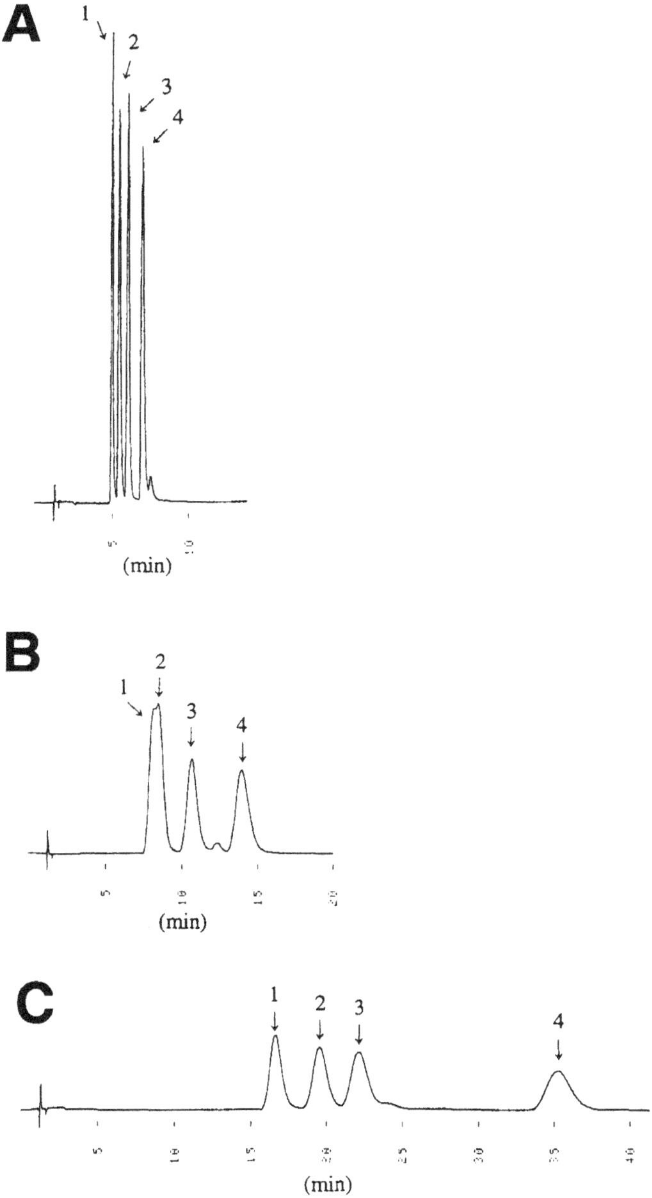

Fig. 3. Effect of column on chiral separations by SFC: A, Chiralcel OD; B, Chiralcel OC; C, Chiralcel OF. SFC conditions: mobile phase CO_2-13%(v/v) isopropanol containing 0.5%(v/v) diethylamine, flow rate of CO_2 2 mL/min, outlet pressure 180 bar, temperature 50°C, detection at 254 nm. Peaks: *1*, *L-trans* isomer; *2*, *D-trans* isomer; *3*, *l-cis* isomer; *4*, *d-cis* isomer. (From **ref. *16*** with permission of Elsevier Science-NL, Amsterdam, The Netherlands.)

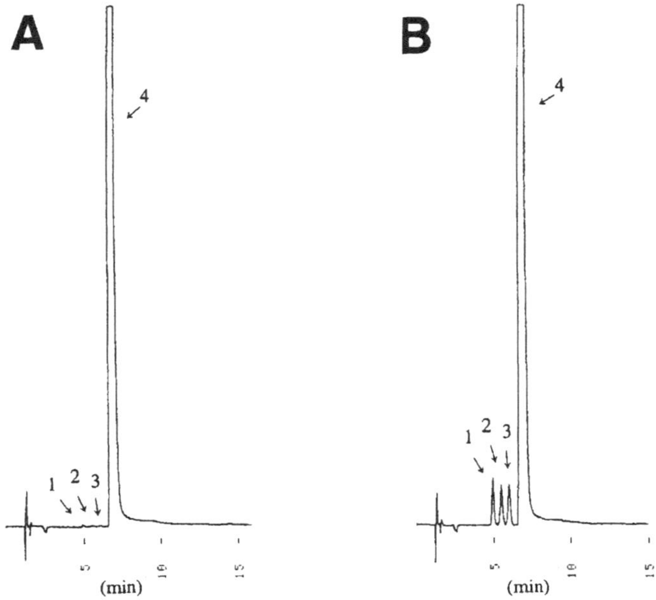

Fig. 4. Chromatograms, using a Chiralcel OD column with SFC conditions and peaks as in **Fig. 3**, of diltiazem spiked with the three optical isomers. **(A)** 0.05%. **(B)** 1.0%. (From **Ref. *16*** with permission of Elsevier Science-NL, Amsterdam, The Netherlands.)

6. In a normal-phase HPLC, little attention has been focused on the column temperature from a practical point of view. This is mainly due to the use of a combustible organic solvent such as *n*-hexane. It is one of the significant merits of p-SFC that noncombustible carbon dioxide is used as the main mobile phase constituent. The remarkable change in density with temperature can be expected because of significant compressibility of supercritical carbon dioxide.
7. The mobile phase is always fed in a constant-flow delivery mode in this system.
8. The effect of diethylamine is an improvement of the peak shape by the deactivation of the active sites on the silica support.
9. The ratio of modifier to carbon dioxide is one of volume and the conditions should be quoted as v/v. The modifier is mixed volumetrically with carbon dioxide by controlling the pumping rates. The system can also be operated in a gradient elution mode by programming the flow-rate of the modifier *(17)*.
10. A 5 µL sample loop should be used, as specified. Use of larger sample loops of 10 or 20 µL produces a broader and/or split peak shape due to the difference of properties between carbon dioxide and the sample solvent.
11. A pressure drop of about 20 bar will be produced through the column.
12. The separation factor, the resolution and the plate number on the Chiralcel OD column are 1.13, 1.65, and 5895 for *trans* enantiomers, respectively, and 1.17, 2.27, and 6137 for *cis* enantiomers, respectively.

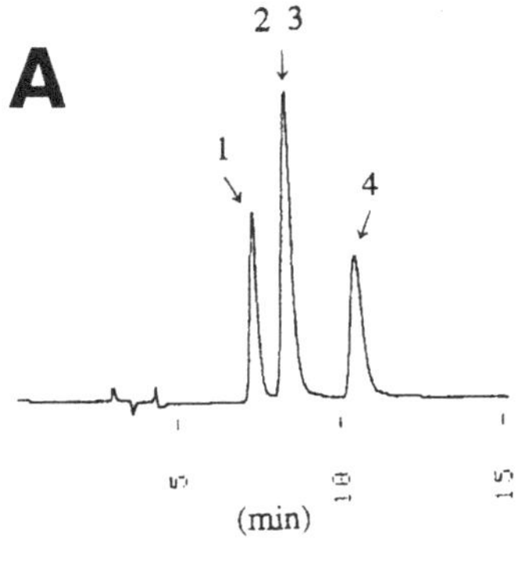

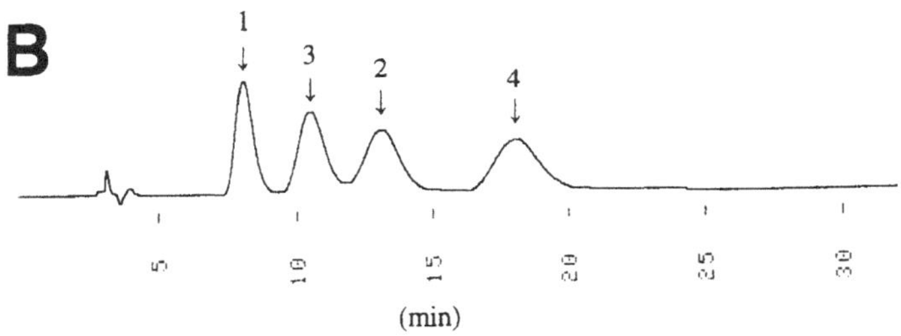

Fig. 5. Chiral separations of diltiazem optical isomers by HPLC with: A, Chiralcel OD; B, Chiralcel OF. HPLC conditions: mobile phase *n*-hexane-isopropanol **(A)** 9:1; **(B)** 1:1 containing 0.1%(v/v) diethylamine; flow rate 1 mL/min; temperature 30°C; detection at 254 nm. Peaks as in **Fig. 3**. (From **ref. *16*** with permission of Elsevier Science-NL, Amsterdam, The Netherlands.)

13. The difference in the retention between diltiazem and its three optical isomers can be explained by the difference in the extent of the following interactions: the interaction between the 4-methoxylphenyl group of the solute and the phenyl group of the chiral stationary phase (CSP); and the interaction by hydrogen bonding between the ester group of the solute and the carbamate group of the CSP.

References

1. Petersson, P. and Markides, K. E. (1994) Chiral separations performed by supercritical fluid chromatography. *J. Chromatogr. A* **666,** 381–394.
2. Lee, C. R., Porziemsky, J.-P., Aubert, M.-C., and Krstulovic, A. M. (1991) Liquid and high-pressure carbon dioxide chromatography of β-blockers: resolution of the enantiomers of nadolol. *J. Chromatogr.* **539,** 55–69.

3. Kot, A., Sandra, P., and Venema, A. (1994) Sub- and supercritical fluid chromatography on packed columns: a versatile tool for the enantioselective separation of basic and acidic drugs. *J. Chromatogr. Sci.* **32,** 439–448.
4. Biermanns, P., Miller, C., Lyon, V., and Wilson, W. (1993) Chiral resolution of β-blockers by packed-column supercritical fluid chromatography. *LC-GC* **11,** 744–747.
5. Siret, L., Macaudiere, P., Bargmann-Leyder, N., Tambuté, A., Caude, M., and Gougeon, E. (1994) Separation of the optical isomers of a new 1,4–dihydropyridine calcium channel blocker (LF 2.0254) by liquid and supercritical fluid chromatography. *Chirality* **6,** 440–445.
6. Anton, K., Eppinger, J., Frederiksen, L., Francolte, E., Berger. T. A., and Wilson. W. H. (1994) Chiral separations by packed-column super- and subcritical fluid chromatography. *J. Chromatogr. A* **666,** 395–401.
7. Wang, Z., Klee, M. S., and Yang, S. K. (1995) Achiral and chiral analysis of camazepam and metabolites by packed-column supercritical fluid chromatography. *J. Chromatogr. B* **665,** 139–146.
8. Stringham, W. (1996) Relationship between resolution and analysis time in chiral subcritical fluid chromatography. *Chirality* **8,** 249–257.
9. Lynam, G. and Nicolas, E. C. (1993) Chiral HPLC versus chiral SFC: evaluation of long-term stability and selectivity of Chiralcel OD using various eluents. *Biomed. Anal.* **11,** 1197–1206.
10. Stringham, W., Lynam, K. G., and Grasso, C. C. (1994) Application of subcritical fluid chromatography to rapid chiral method development. *Anal. Chem.* **66,** 1949–1954.
11. Shimizu, R., lshii, K., Tsumagari, N., Tanigawa, M., and Matsumoto, M. (1982) Determination of optical isomers in diltiazem hydrochloride by high-performance liquid chromatography. *J. Chromatogr.* **253,** 101–108.
12. Shimizu, R., Kakimoto, T., lshii, K., Fujimoto, Y., Nishi, H., and Tsumagari, N. (1986) New derivatization reagent for the resolution of optical isomers in diltiazem hydrochloride by high-performance liquid chromatography. *J. Chromatogr.* **357,** 119–125.
13. Ishii, K., Banno, K., Miyamoto, T., and Kakimoto, T. (1991) Determination of diltiazem hydrochloride enantiomers in dog plasma using chiral stationary-phase liquid chromatography. *J. Chromatogr.* **564,** 338–345.
14. Nishi, H., Fujimura, N., Yamaguchi, H., and Fukuyama, T. (1993) Direct high-performance liquid chromatographic separation of the enantiomers of diltiazem hydrochloride and its 8-chloro derivative on a chiral ovomucoid column. *J. Chromatogr.* **633,** 89–96.
15. Ishii, K., Minato, K., Nishimura, N., Miyamoio, T., and Sato, T. (1994) Direct chromatographic resolution of four optical isomers of diltiazem hydrochloride on a Chiralcel OF column. *J. Chromatogr. A* **686,** 93–100.
16. Yaku, K., Aoe, K., Nishimura, N., Sato, T., and Morishita, F. (1997) Chiral resolution of four optical isomers of diltiazem hydrochloride on Chiralcel columns by packed-column supercritical fluid chromatography. *J. Chromatogr. A* **785,** 185–193.
17. Yaku, K., Aoe, K., Nishimura, N., Sato, T., and Morishita, F. (1997) Retention behavior of synthetic corticosteroids in packed-column supercritical fluid chromatography. *J. Chromatogr. A* **773,** 277–284.

20

Determination of Salbutamol Sulfate and Its Impurities in Pharmaceuticals by Supercritical Fluid Chromatography

María J. del Nozal, Laura Toribio, José L. Bernal, and María L. Serna

1. Introduction

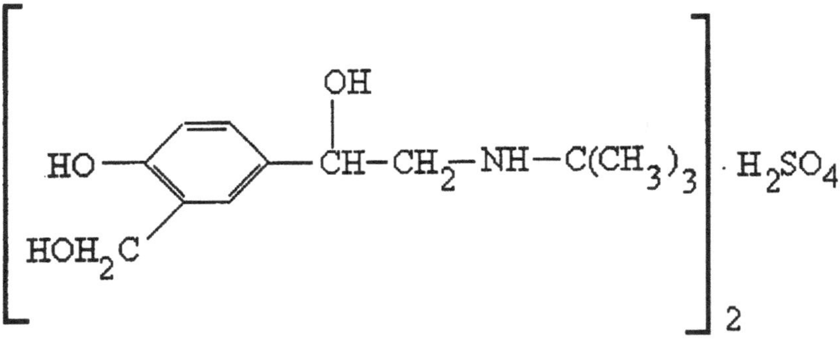

Salbutamol sulfate, shown above, is a bronchodilator used for the treatment of asthma. Most of the papers published in relation to salbutamol sulfate analysis described its determination and quantification in tissues and biological fluids of animals under treatment with this drug. Normally, the methods employed are based on high-performance liquid chromatography (HPLC) techniques using detectors of high sensitivity such as fluorescence *(1–3)* and electrochemical *(4–6)*. It is known that there are some impurities that could be produced during synthesis or during storage of the drug. Consequently, there is great interest in the analysis of the drug and its impurities. HPLC *(7)* or capillary electrophoresis *(8)* methods have been used to determine

From: *Methods in Biotechnology, Vol. 13: Supercritical Fluid Methods and Protocols*
Edited by: J. R. Williams and A. A. Clifford © Humana Press Inc., Totowa, NJ

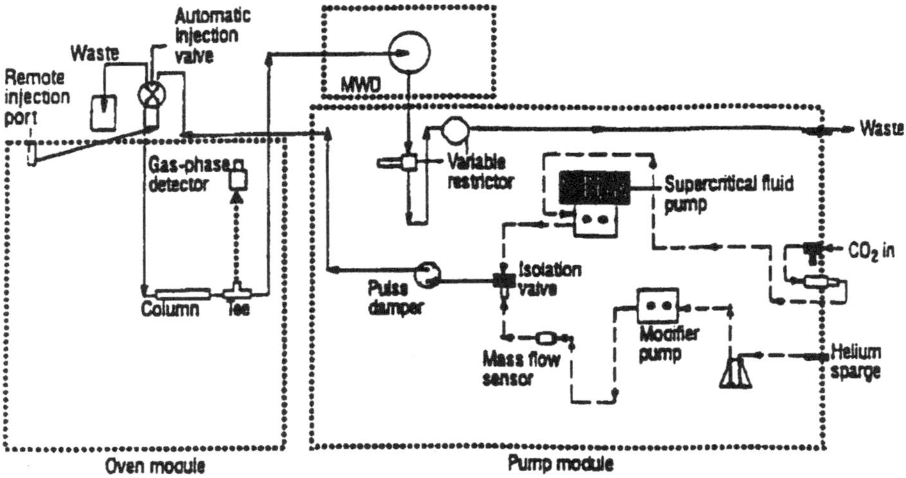

Fig. 1. HP experimental system used.

these compounds, but only two of the impurities were analyzed. Recently, salbutamol sulfate and six related impurities were separated by HPLC in 30 min *(9)*. To shorten this time, analysis of the same samples by supercritical fluid chromatography (SFC) was tried. It was found that analysis times of less than 15 min were possible *(10)*.

This chapter describes a method that allows rapid determination of salbutamol sulfate and six of its related impurities—5-formylsaligenin, salbutamol ketone, salbutamol bisether, isopropylsalbutamol, desoxysalbutamol sulfate, and salbutamol aldehyde—in pharmaceuticals by using packed column SFC with diode array detection (*see* **Fig. 1**).

2. Materials

1. A Hewlett-Packard G1205A supercritical fluid chromatograph (Palo Alto, CA) with an HP1050 diode array detector, an HP7673 GC/SFC autosampler and a Rheodyne (Cotati, CA) valve (5-µL loop). Chromatographic data are collected by means of an HP-SFC 3365 Chemstation.
2. A 5-µm Lichrospher Diol column, 250 mm × 4.6 mm, from Phenomenex (Torrance, CA).
3. *N*-Propylamine, dimethylamine, and *n*-butylamine are purchased from Sigma Aldrich Química (Madrid, Spain). Methanol (HPLC-grade) is obtained from Lab-Scan (Dublin, Ireland). Samples and drug-certified standards are kindly supplied by Glaxo-Wellcome S.A. (Aranda de Duero factory, Burgos, Spain).
4. Carbon dioxide (minimum purity 99.999%), kept in cylinders with a diptube, and supplied by Air Products (Sombreffe, Belgium), is used in all the experiments as the mobile phase.

5. Ultrasonic bath from Selecta (Barcelona, Spain).
6. Vibromatic shaker from Selecta (Barcelona).
7. Centrifuge 5415C from Eppendorf (Hamburg, Germany).
8. Ultrapure water is obtained from a Milli-Q apparatus from Millipore (Bedford, MA).
9. Pipettes, volumetric flasks, and other common glassware are also employed.

3. Method

3.1. Preparation of Standard Solutions

1. Weigh 10 mg of compound (*see* **Notes 1** and **2**), transfer with methanol to a volumetric flask of 10 mL. Dissolve and complete the volume with a 1:1 mixture of water:methanol (*see* **Note 3**). Repeat this operation with all the compounds to be analyzed (*see* **Note 4**).
2. Make dilutions (with a 1:1 water:methanol mixture) of all the solutions to get at least seven points on a calibration plot. The range of concentrations should cover 1 to 10 μg/mL (*see* **Note 5**).

3.2. Sample Preparation

3.2.1. Tablets

1. Take at random 5 tablets and grind them in a glass mortar (*see* **Note 6**).
2. Transfer all the powder to a volumetric flask of 100 mL.
3. Add 50 mL of ultrapure water and 0.1 mL of 12 *N* HCl.
4. Shake mechanically for 3 min.
5. Sonicate for 30 s.
6. Shake mechanically for 1 h.
7. Complete the volume (100 mL) with ultrapure water.
8. Leave to stand for 10 min.
9. Take an aliquot of 10 mL.
10. Centrifuge at 3400 g for 10 min.
11. Take a portion of the liquid phase and fill a 2-mL topaz vial ready for analysis.

3.2.2. Syrups

1. Transfer with a pipette 1 mL of sample to a 50 mL volumetric flask (*see* **Note 6**).
2. Add 10 mL of methanol.
3. Sonicate for 5 min.
4. Complete the volume with ultrapure water.
5. Take an aliquot in a topaz vial of 2 mL.

3.2.3. Placebo

1. Mix all the excipients in the same proportion as in the formulation, but without adding the compounds to be analyzed.
2. Use aliquots of this mixture to test its influence on the determination of the different compounds and also to know the blank average signal.

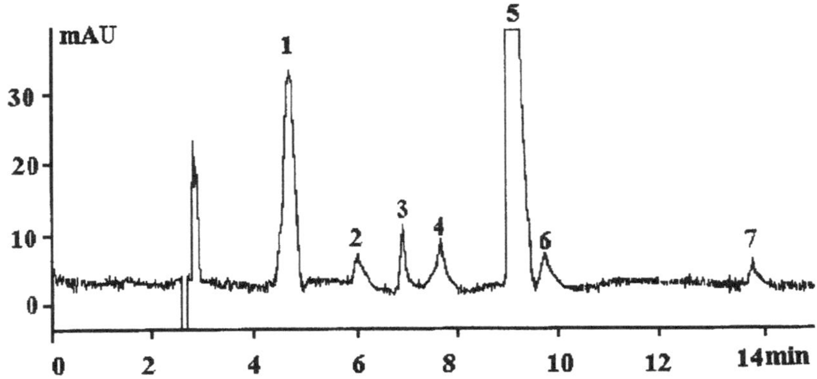

Fig. 2. Chromatogram of a mixture of standards : *1,* 5-formylsaligenin 10 µg mL^{-1}; *2,* salbutamol ketone 10 µg mL^{-1}; *3,* desoxysalbutamol 10 µg mL^{-1}; *4,* salbutamol aldehyde 10 µg mL^{-1}; *5,* salbutamol sulfate 500 µg mL^{-1}; *6,* isopropylsalbutamol 10 µg mL^{-1}; *7,* salbutamol bisether 30 µg mL^{-1}.

3.3. SFC Analysis

1. The instrument is operated in the downstream mode. Pressure and temperature are fixed at 300 bar and 70°C, respectively (*see* **Note 7**). The flow rate is 1.5 mL/min (*see* **Note 8**) and a gradient (*see* **Note 9**) of modifier [methanol with 0.5% *n*-propylamine (*see* **Notes 10** and **11**)] is used. The injection volume is 5 µL (full loop). Inject the standard calibration solutions into the SFC system (*see* **Note 12**). A sample chromatogram is shown in **Fig. 2**.
2. Integrate the chromatograms and report the peak area of the different compounds.
3. Verify linearity over the range selected. Correlation coefficients must be better than 0.99.
4. Verify the limits of detection and quantitation (LOD and LOQ) established by the equations:

$$LOD = 3 \times \sigma_{x/y}/b,$$

$$LOQ = 10 \times \sigma_{x/y}/b.$$

   where $\sigma_{x/y}$ is the standard deviation of the linear fitting and b is the slope of the fitting. The detection limits are usually ranged between 0.20 to 0.50 µg/mL, except for salbutamol bisether for which the detection limit is 1.30 µg/mL.
5. Check recovery of all the compounds, the interday repeatability and the intraday reproducibility on placebo-spiked samples (*see* **Note 13**).
6. Make blanks applying the SFC method to placebo samples to test for the presence of possible interfering peaks from the matrix (*see* **Note 14**).
7. Recoveries of all the compounds must be higher than 95%. Repeatabilities and reproducibilities, as measured by relative standard deviation, must be better than 2.5% and 4%, respectively (ten determinations).

4. Notes

1. The solid standards must be kept at temperatures lower than 4°C (in the refrigerator), in well-sealed vials and protected from light. It is convenient to put the vials into another container where desiccant has previously been added. Following these recommendations, the standards can last at least 1 yr.
2. When the samples are exposed to extraordinary storage conditions (temperatures between 35 and 80°C, and relative humidity from 60% to 100% for periods of up to 1 yr), large quantities of degradation products appear. In these extreme conditions, and mainly when the temperature is higher, it is a problem for the evaluation of isopropylsalbutamol. More specifically, a bigger isopropylsalbutamol peak that easily overlapped with the corresponding peak for saccharin was encountered.
3. It is advisable to prepare small quantities of stock solutions, protect them from light, keep them in the refrigerator and replace them every month.
4. Some compounds are supplied in very small quantities in sealed vials, so it is more convenient to dissolve all the contents *in situ* with methanol and then dilute to the required concentration.
5. For the calibration of the bisether, it is advisable to make the concentration range of the order of 3 to 30 μg/mL. The reasons are that its response is lower and, moreover, it is the last one to elute in the zone where the baseline starts to grow.
6. Under normal storage conditions, neither the syrups nor the tablets present problems in the determination of salbutamol sulfate. For the samples that Glaxo-Wellcome supplied, any impurities at a level higher than the detection limits are undesirable.
7. If the working pressure is reduced, significant changes in the resolution between salbutamol aldehyde and desoxyalbutamol can be expected. The resolution changes from 0.51 at 150 bar to 1.75 at 300 bar. Generally, the oven temperature has little influence on retention and selectivity. However, it must be taken into account that near 60°C there is a change in the elution order of salbutamol ketone, salbutamol aldehyde, and desoxysalbutamol.
8. If a flow rate of 1 mL/min is used, then the retention times are increased by about 50%.
9. The most adequate organic modifier gradient profile was initially 30%, held for 9.5 min and then programmed to increase at 1.5%/min to 45%.
10. The addition of an amine to the modifier enhances the peak shape and also reduces the retention time of all the compounds.
11. Of the amines considered (*n*-propylamine, dimethylamine, and *n*-butylamine), propylamine gives the best results up to a concentration of 0.5%. However, adding more than 0.5% did not improve the chromatography significantly.
12. Using the described conditions, the retention times are 4.9 min (5-formylsaligenin), 6.1 min (salbutamol ketone), 7.1 min (desoxysalbutamol), 7.8 min (salbutamol aldehyde), 9.2 min (salbutamol ketone), 9.8 min (isopropylsalbutamol), and 14.4 min (salbutamol bisether).
13. The method allows the determination of different compounds in ratios of up to 1000:1 (salbutamol sulfate:impurity), and these are near the levels expected in

real cases. In this situation, it is also convenient to prepare placebo-spiked aliquots in the same proportions.

14. When cough syrup samples are analyzed, some problems appear due to the excipient, 5-formylsaligenin coeluting with benzoate and isopropylsalbutamol coeluting with saccharin. This shortcoming could be circumvented by either using an alternative HPLC method or by introducing a clean-up stage in the procedure. The latter step would complicate the sample treatment and implies a longer analysis time.

References

1. Degroodt, J. M., Debukanski, B. W., and Srebrnik, S. (1992) Immunoaffinity-chromatography purification of salbutamol in liver and HPLC-fluorometric detection at trace residue level. *Z. Lebens* **195,** 566–568.
2. McCarthy, P. T., Atwal, S., Sykes, A. P., and Ayres, J. G. (1993) Measurement of terbutaline and salbutamol in plasma by high performance liquid chromatography with fluorescence detection. *J. Biomed. Chromatogr.* **7,** 25–28.
3. Gupta, R. N., Fuller, H. D., and Dolovich, M. B. (1994) Optimization of a column liquid chromatographic procedure for the determination of plasma salbutamol concentration. *J. Chromatogr. B* **654,** 205–211.
4. Sagar, K. A., Hua, C., Kelly, M. T., and Smyth, M. R. (1992) Analysis of salbutamol in human plasma by high performance liquid chromatography with electrochemical detection using a micro electrochemical flow cell. *Electroanalysis* **4,** 481–486.
5. Sagar, K. A., Kelly, M. T., and Smyth, M. T. (1993) Simultaneous determination of salbutamol and terbutaline at overdose levels in human plasma by high performance liquid chromatography with electrochemical detection. *J. Biomed. Chromatogr.* **7,** 29–33.
6. Ramos, F., Castihlo, M. C., Dasilveira, M. I. N., Prates, J. A. M., and Correira, J. H. R. (1993) Determination of salbutamol in rats at low concentrations using liquid chromatography with electrochemical detection. *Anal. Chim. Acta* **275,** 279–283.
7. Mulholland, M. and Waterhouse, J. (1988) Investigation of the limitations of saturated fractional factorial experimental designs, with confounding effects for an HPLC ruggedness test. *Chromatographia* **25,** 769–774.
8. Altria, K. D. (1993) Determination of salbutamol related impurities by capillary electrophoresis. *J. Chromatogr.* **634,** 323–328.
9. Bernal, J. L., Nozal, M^a. J., Velasco, H., and Toribio, L. (1996) HPLC versus SFC for the determination of salbutamol sulphate and its impurities in pharmaceuticals. *J. Liq. Chrom. Rel. Technol.* **19,** 1579–1589.
10. Bernal, J. L., Nozal, M^a. J., Rivera, J. M., Serna, M^a. L., and Toribio, L. (1996) Separation of salbutamol and six related impurities by packed column supercritical fluid chromatography. *Chromatographia* **42,** 89–94.

21

Packed Column Supercritical Fluid Chromatographic Determination of Acetaminophen, Propyphenazone, and Caffeine in Pharmaceutical Dosage Forms

Urmila J. Dhorda, Viddesh R. Bari, and M. Sundaresan

1. Introduction

Supercritical fluid chromatography (SFC), particularly with packed columns, has recently being gaining in popularity and is being investigated with increasing frequency for the characterization of pharmaceutical and biological agents. SFC can be described, roughly, as a form of high-performance liquid chromatography (HPLC), in which a fluid kept above its critical pressure and temperature, replaces the liquid-mixture mobile phase, which is normally used in HPLC. As the majority of drugs are either polar or moderately polar, pure supercritical carbon dioxide, being nonpolar, is not applicable to pharmaceutical analysis. This difficulty can be easily overcome by the use of a two-component mobile phase consisting of supercritical carbon dioxide and a small amount of a polar solvent. The increased solvent strength of this two-component mobile phase can be attributed to dipole–dipole, dipole-induced dipole, dispersive and hydrogen bonding (acidic and basic) forces. This mobile phase can solvate most known drugs and thus becomes a versatile mobile phase. The polar, organic solvent is known as the modifier, and modifiers used include methanol, ethanol, isopropanol, dichloromethane, tetrahydrofuran, dimethyl sulfoxide, and acetonitrile. The polar nature of this mobile phase can further be tailored to achieve retention by the addition of smaller quantities of weak acids or bases like trimethylamine, formic acid, acetic acid, and so on.

A wide number of applications of SFC to drug and pharmaceutical analysis have been published. Berger and Wilson *(1)* have demonstrated the technique

From: *Methods in Biotechnology, Vol. 13: Supercritical Fluid Methods and Protocols*
Edited by: J. R. Williams and A. A. Clifford © Humana Press Inc., Totowa, NJ

for the rapid separation of 10 phenothiazine antipsychotics in 11 min, 10 tricyclic antidepressant drugs in less than 6 min *(2)*, and 9 stimulants in 15 min *(3)*, using a three-component mobile phase consisting of supercritical carbon dioxide, methanol, and isopropylamine. They used absorptimetric detection and showed that packed column SFC offered a viable means to separate many drugs. Even though detection limits were similar to those obtained by liquid chromatography, this technique was faster and more efficient than liquid chromatography. Strode et al. *(4)* extended the application of this technique to the determination of felodipine. Major uses of this technique have also been demonstrated for chiral chromatography. Misoprostol, a prostaglandin, was determined from 200-μg tablets by combined supercritical fluid extraction and SFC by Patel, Dhorda, and Sundaresan *(5)* using this technique. The versatility of this technique was demonstrated by Bhoir et al. *(6)* who separated and quantified seven vasodilators belonging to different families. Bari, Dhorda, and Sundaresan *(7)* determined acetaminophen, chlorzoxazone and ibuprofen by modifier flow programming.

A fast and efficient, isocratic, isobaric, and isothermal protocol is presented in this chapter for the packed column SFC separation and quantitation of three useful drugs from a combined dosage form using an internal standard method. These are acetaminophen (*N*-acetyl-*p*-aminophenol), propyphenazone (4-isopropyl-2, 3-dimethyl-1-phenyl-3-pyrazolin), and caffeine.

2. Materials

1. A supercritical fluid chromatograph configured with two pumps for dynamic mixing of carbon dioxide and methanol and with flow rate adjustments for both (0.01–10 mL/min). Outlet pressure programming should be available from 7.38 to 35.0 MPa and temperature programming from 35°C to 80°C. It should have a Rheodyne injector with a 20 μL external loop. The chromatograph should have a multiwavelength spectrophotometric detector (190–600 nm) and 5 mm pathlength, 4 μL high-pressure flow cell connected to a Borwin software integrator and printer.
2. A 250 × 4.6 mm column for reverse-phase SFC, i.e., with an octadecyl (C_{18}) bonded silica 10 μm packing.
3. A microliter syringe of capacity 25 μL, e.g., from Hamilton.
4. Methanol, which should be HPLC grade, filtered through a 0.45-μm filter to remove all particulate matter, degassed using an ultrasonic bath sonicator and stored in reservoir.
5. Standard samples of acetaminophen, propyphenazone, caffeine, and ibuprofen with certificates of assay.
6. Mobile phase waste collector.

3. Method

1. Prepare separate stock solutions of acetaminophen, prophyphenazone, caffeine and ibuprofen by weighing 100 mg of each drug and dissolving in 100 mL of methanol.

2. Condition the equipment by switching on the chromatograph and allow it to warm up for 10 min.
3. Degas the methanol, keep it in a reservoir and connect the reservoir to the chromatograph.
4. Set the pressure switch on the SFC chromatograph at 12.75 MPa and the temperature at 45°C. Set the flow rate of CO_2 at 2.0 mL/min and modifier at 0.1 mL/min (*see* **Note 1**).
5. Fix the C_{18} column between the injector and the detector.
6. Open the gas valve and the modifier valve and allow the mixture to flow through the system and column for 10 min to condition the apparatus.
7. Prepare six mixture solutions in methanol of volume 10 mL containing 10.0, 20.0, 30.0, 50.0, 80.0, and 100.0 μg/mL of the three drugs and 50 μg/mL of the ibuprofen internal standard.
8. Set the detector at 230 nm.
9. Inject 20 μL of each of the above solutions, starting from the lower concentration. Obtain chromatograms and measure responses as peak heights (*see* **Note 2**). Calculate detector responses as peak height ratios of the drug/internal standard.
10. Plot calibration graphs and calculate slope and intercept by the linear regression (least-squares fit) method. Calculate also the standard deviations in slope and intercept, correlation factor, and point error (*see* **Notes 3–6**).
11. Obtain the mean weight of 20 tablets containing the three drugs. Crush the tablets, obtain a fine powdery form and homogenize. Weigh out a portion of this powder equivalent to the mean weight of a tablet. Dissolve this portion in 100 mL of methanol with stirring and filter. Dilute an appropriate aliquot of this solution with methanol to bring the solution in the range of 10–100 μg/mL for the three drugs. Add an appropriate aliquot of the stock solution of ibuprofen to give 50 μg/mL of the internal standard and make up to volume. With the chromatograph set to the parameters given above, inject 20 μL of the solution. Measure peak heights and calculate the peak heights ratio of the drug/internal standard. Use the calibration data obtained in **step 10** to obtain the concentrations in the tablets. Repeat this experiment seven times, find mean values, and compare these values with the labeled amounts in the tablets (*see* **Note 7**).

4. Notes

1. Adjustment of the optimum parameters given above may produce improved chromatograms.
2. A typical chromatogram of the separation of the three drugs and the internal standard is given in **Fig. 1**. The chromatographic conditions were somewhat different from the optimum given above and were; pressure 9.81 MPa; temperature 40°C, rate of flow of CO_2 2.0 mL/min and rate of flow of methanol 0.15 mL/min.
3. An example of the calibration data obtained is given in **Table 1**.
4. The lowest quantifiable limit was found to be 10 μg/mL. Translated into the actual amount injected this will be 200 ng when 20 μL is injected. When required, lower limits can be obtained by modification of the method. These limits could further

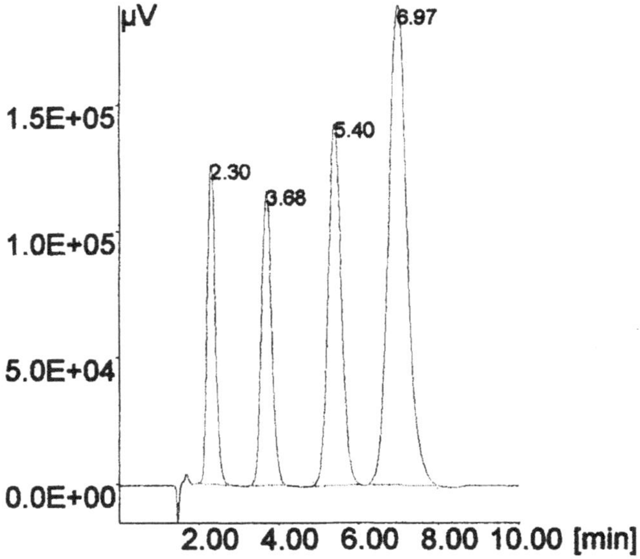

Fig. 1. SFC chromatogram of a mixture of acetaminophen, propyphenazone, caffeine and ibuprofen obtained with the following conditions: pressure 9.81 MPa; temperature 40°C, rate of flow of CO_2 2.0 mL/min and rate of flow of methanol 0.15 mL/min.

Table 1
Linear Regression (Least-Squares Fit) Calibration Data

	Acetaminophen	Propyphenazone	Caffeine
Concentration range (μg/mL)	10-100	10-100	10-100
Slope m	0.0256	0.0218	0.0302
Intercept b	0.0070	0.0020	0.0111
SD of slope S_m	0.0004	0.0009	0.0025
SD of intercept S_b	0.0254	0.0541	0.1466
Correlation coefficient r	0.9999	0.9999	0.9999
Point error S_{yx}	0.0346	0.0736	0.1997

be reduced by the appropriate choice of wavelength if only one or two of the drugs are to be determined. The present choice of wavelength of 230 nm is the compromise value for all the four drugs.

5. Recovery experiments showed that the average recovery was 99.5 ± 0.2% in the high ranges and 95.2 ± 0.5% in the lower ranges.
6. Results of the intraday and interday performance experiments showed that packed column SFC was highly reproducible, and the relative standard deviation during these periods never exceeded 5%.

7. In an example sevenfold replicate analysis, the content of acetaminophen tablets was found to be 250.2 ± 0.4 mg propyphenazone, 149.8 ± 0.3 mg and 49.9 ± 0.5 mg of caffeine against the labeled amounts of 250 mg, 150 mg, and 50 mg, respectively.

References

1. Berger, T. A. and Wilson, W. H. (1994) Separation of drugs by packed column supercritical fluid chromatography. 1. Phenothiazine antipsycotics. *J. Pharm. Sci.* **83,** 281–286.
2. Berger, T. A. and Wilson, W. H. (1994) Separation of drugs by packed column supercritical fluid chromatography. 2. Antidepressants. *J. Pharma. Sci.* **84,** 287–290.
3. Berger, T. A. and Wilson, W. H. (1995) Separation of basic drugs by packed column supercritical fluid chromatography. 3. Stimulants. *J. Pharm. Sci.* **84,** 489–492.
4. Strode, III J. T. B., Taylor, L. T., Howard, A. L., Ip. D., and Brooks, M. A. (1994) Analysis of felodipine by packed column supercritical fluid chromatography with electron capture and ultraviolet absorbance detection. *J. Pharm. Biomed. Anal.* **12,** 1003–1014.
5. Patel, Y. P., Sundaresan, M., and Dhorda, U. J. (1997) Supercritical fluid extraction and chromatography of misoprostol from tablets. *Ind. J. Pharm. Sci.* **59,** 132–134.
6. Bhoir, I. C., Raman, B., Sundaresan, M., and Bhagwat, A. M. (1998) Separation and estimation of seven vasodilators using packed column supercritical fluid chromatography. *J. Pharm. Biomed. Anal.* **17,** 539.
7. Bari, V. R., Dhorda, U. J., and Sundaresan, M., (1997) A simultaneous packed column supercritical fluid chromatographic method for ibuprofen, chlorzoxazone and acetaminophen in bulk and dosage forms. *Talanta* **45,** 297–302.

22

Analysis of Shark Liver Oil by Thin-Layer and Supercritical Fluid Chromatography

Christina Borch-Jensen,
Magnus Magnussen, and Jørgen Mollerup

1. Introduction

The liver oils of certain shark species contains squalene, 2,6,20,15,19,23-hexamethyltetracosahexane, at high levels. Squalene is used in the pharmaceutical, rubber, and surfactants industries *(1)*. Squalene is easily hydrogenated to give squalane, which is an important raw material in the cosmetic industry where it is used as a skin lubricant and in the pharmaceutical industry where it is used as a carrier for fat-soluble drugs *(1)*. The price of shark liver oil for these purposes is determined from the squalene content of the oil, and therefore reliable methods for the determination of squalene are necessary. Supercritical fluid chromatography or SFC (*see* Chapter 1) is a well-suited method for the analysis of underivatized marine oils *(2)*, and determination of the squalene content can be done with a minimum of sample preparation *(3)*. A more time-consuming method, determination of iodine value according to the AOAC standard method, can be applied for a rough estimate of the content of squalene in shark liver oils. The iodine value is a measure of unsaturation in the oil and the high degree of unsaturation of the fatty acids in shark liver oil makes it difficult to distinguish between the kind of components that contribute to the iodine value. However, it has been shown that a linear relationship between iodine value and squalene content found by SFC analysis exists *(3)*.

Besides squalene, shark liver oils may contain high levels of the so-called ether lipids (diacylglycerol ethers or 1-alkyl-2,3-diacylglycerols) (*see* **Fig. 1**). These are of interest because of their similarities to the plate activating factors (PAF) *(4)*. As seen from **Fig. 1** the ether lipids and the triglycerides differ in the way that the ether lipids have the fatty acids in position 1 linked by an ether

From: *Methods in Biotechnology, Vol. 13: Supercritical Fluid Methods and Protocols*
Edited by: J. R. Williams and A. A. Clifford © Humana Press Inc., Totowa, NJ

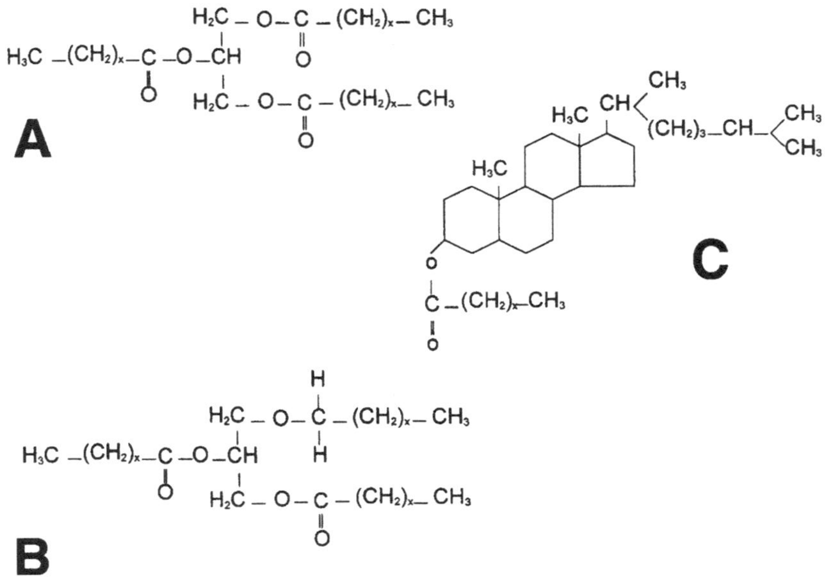

Fig. 1. The structures of **(A)** triglycerides, **(B)** ether lipids, and **(C)** cholesterol esters.

bond while this bond is an ester bond in a triglyceride. Ether lipids in shark liver oils have molecular weights very close to those of triglycerides, and a high degree of unsaturation. This makes analysis by gas chromatography (GC) very difficult because the large polyunsaturated molecules can not withstand the high temperatures without polymerization.

SFC has the advantage of lower analysis temperature and is therefore suited to the analysis of polyunsaturated triglycerides and ether lipids. However, when analyzing raw shark liver oils on a nonpolar capillary column there will be coelution between ether lipids and triglycerides because of the almost similar structure and molecular weights of these two lipid groups (*see* **Fig. 1**). The two lipid groups have a small difference in polarity, the ether lipids being less polar because of the one ether bond. This difference is difficult to make use of using a nonpolar SFC column and a nonpolar (CO_2) mobile phase. To complicate matters further, shark liver oils have a rather high content of cholesterol esters, which will also elute in this area of the chromatogram.

By thin layer chromatography (TLC) on polar silica plates it is possible to separate the triglycerides from the ether lipids and the cholesterol esters using a nonpolar mobile phase. The TLC method can be scaled up to yield fractions with enough sample for further analysis by SFC.

This protocol describes a method for a detailed analysis of shark liver oil lipids by TLC and capillary SFC. The shark liver oil is fractionated by preparative TLC and the fractions are analyzed by SFC. This gives information on the overall composition of the oil and on the carbon number distribution within the single lipid groups.

2. Materials

1. Solvents: petrol ether, diethyl ether, acetic acid, chloroform, *n*-heptane, isooctane, or cyclohexane (*see* **Note 1**).
2. Gases: carbon dioxide, helium, atmospheric air, hydrogen, nitrogen (*see* **Note 2**).
3. TLC tank for 20 × 20 cm plates.
4. TLC plates: silica 60 in sizes 20 × 20 cm and 5 × 20 cm. Thickness of coating: 0.25 cm (*see* **Note 3**).
5. 100-µL syringe for sample application.
6. Apparatus for evaporation of solvents (*see* **Note 4**).
7. Spraying equipment for TLC plates.
8. TLC plate heater.
9. Glassware, including 30 mL screw cap vials, Pasteur pipettes.
10. SFC instrument featuring capillary column operation.
11. 20-m SFC capillary column with an inner diameter of 0.2 mm and stationary phase of 5% phenyl-95% methylsilicone.

3. Method

1. Prepare the TLC mobile phase and tank. Petrol ether, diethyl ether, and acetic acid are mixed in the ratio 85:15:1.5 by volume, respectively. The mobile phase is poured into the tank and left for a minimum of 1 h to ensure complete saturation of the tank (*see* **Note 5**).
2. Apply the shark liver oil to the TLC plates as follows. The melted shark liver oil (*see* **Note 6**) is dissolved in *n*-heptane, isooctane, or cyclohexane to a concentration of 1 g/mL. The sample is applied to the TLC plate 1 cm from the bottom as a 20-cm band. An even distribution of the sample is essential and is easily accomplished by the use of a syringe with a 90° tip. As much as 75 µL can be applied to a 20 × 20 cm plate corresponding to a load of 3.75 mg/cm plate. A similar loading is also applied to a smaller 5 × 20 cm plate. Two 20 × 20-cm plates are used to ensure enough sample output for the SFC analysis, together with the smaller 5 × 20 cm plate for monitoring the separation (*see* **Note** 7).
3. Develop the plates as follows. The plates are placed in the TLC tank and left for development until the solvent front is 1 cm from the top of the plate. This will take approximately 45 min. After development, the plates are left to dry for 5 min. The small plate is sprayed with a 5% sulfuric acid in methanol solution (*see* **Note 8**) and left to dry before heating to 120°C on the TLC plate heater to visualize the bands. **Table 1** gives the approximate retention factors relative to the solvent front (*see* **Note 9**).
4. Identify the bands of triglycerides, ether lipids, and cholesterol esters and squalene on the small plate. The bands in similar positions are marked on the

Table 1
Retention Factors
for Shark Liver Oil Lipid Groups

Cholesterol	0.1
Free fatty acids	0.19–0.24
Triglycerides	0.27–0.5
Ether lipids	0.53–0.63
Squalene + cholesterol esters	0.85–0.93

large plates and the silica material of each band is scraped completely off the plate and filled into separate screw cap vials.

5. An internal standard solution of wax ester palmityl palmitate (C16:0-C16:0) in *n*-heptane at a concentration of 6.6 mg/mL is prepared; 0.5 mL of the internal standard solution is added to each fraction of scraped-off silica.
6. Recover the lipids in each of the fractions by adding 5 mL of chloroform to the silica material, shake the well-capped vial, and remove the solvent with a Pasteur pipette. This is followed by extraction with two times 5 mL of cyclohexane using the same procedure. Transfer the silica finally to a paper filter by means of 2 mL of cyclohexane and wash the silica with 2 times 1 mL of cyclohexane (*see* **Note 10**).
7. Evaporate the combined solvents at 60°C under a stream of nitrogen.
8. Take up the residue in *n*-heptane to produce a solution for analysis by SFC. 0.25 mL of *n*-heptane is used per 20 × 20 cm plate used. The concentration of sample thus obtained is for a SFC system with an injection valve loop of 1 µL and a flow split ratio of 1:100 (column:waste).
9. Carry out the analysis by SFC. The analysis parameters are as follows. The column temperature is 170°C. The flame ionization detector (FID) is heated to 350°C. A frit restrictor is used to maintain the flow rate of 1 mL/min. The density is programmed from 0.3 g/mL to 0.452 g/mL at a rate of 0.004 g/mL/min. The density is kept at 0.452 g/mL for 16 min, and is then raised to 0.52 g/mL at a rate of 0.001 g/mL/min.
10. Carry out SFC calibrations using the same experimental conditions and appropriate solutions of the compounds of interest and the internal standard (wax ester palmityl palmitate) to obtain relative response factors.
11. Convert integrated areas from the partial SFC chromatograms into masses by means of theoretical response factors ***(5)***.
12. Carry out SFC analysis of the intact oil (before fractionation) and use this as a check on the squalene content found by this method.

4. Notes

1. All solvents should be analytical grade. Petrol ether should have a boiling point of <50°C.
2. The gases should have the following purities: carbon dioxide (N45): 99.995%, helium (N45): 99.995%, atmospheric air: water content <25 ppv, hydrogen (N30): 99.9%, nitrogen (N45): 99.995%.

3. The large plates are for the fractionation, the small plates for determination of the actual retention factors. Prescored plates are available as an alternative, but should be avoided as the solvent front tends to move faster at the edges of the plates resulting in higher retention factors at the edges than at the middle of the plate.
4. The evaporation apparatus should be designed to evaporate organic solvents by heating and feature the possibility of evaporation under an inert atmosphere (nitrogen).
5. The TLC mobile phase should be kept in a screw cap bottle to prevent evaporation and thereby changes in the composition.
6. The shark liver oils should be stored frozen below –18°C. The oil should be completely melted before taking samples for TLC. The liver oils are fluid at room temperature.
7. Two or more 20 × 20 cm plates are required to ensure enough sample in each fraction for SFC analysis. TLC plates with a larger coating thickness are available. Such plates offer a higher loadability.
8. Spraying of the plates with the sulfuric acid solution should be done in a fume hood.
9. The retention factors will vary from day to day and, therefore, a small plate should always be analyzed simultaneously to identify the location of the different bands.
10. Cyclohexane as an extraction solvent can be replaced by *n*-heptane or isooctane. The use of chloroform in the first extraction step will prevent the formation of emulsions. If an emulsion is formed when cyclohexane is added, it can be broken by the addition of salt water or by centrifugation.

References

1. Merck (1976) Squalene, in *Merck Index*, Merck, Damrstadt, Germany.
2. Borch-Jensen, C. and Mollerup, J. (1996) Supercritical fluid chromatography of fish, shark and seal oils. *Chromatographia* **42,** 252–258.
3. Borch-Jensen, C., Magnussen, M. P., and Mollerup, J. (1997) Capillary supercritical fluid chromatographic analysis of shark liver oils. *J. Am. Oil Chem. Soc.* **74,** 497–503.
4. Mangold, H. K., and Palthauf, F. (1983) Ether lipids, in *Biochemical and Biomedical Aspects*, Academic Press, New York.
5. Novák, J. (1975) *Quantitative Analysis by Gas Chromatography*, Marcel Dekker, New York.

23

Enzymatically Catalyzed Transesterifications in Supercritical Carbon Dioxide

Rolf Marr, Harald Michor, Thomas Gamse, and Helmut Schwab

1. Introduction

Supercritical fluids and, in particular, supercritical carbon dioxide ($SCCO_2$) are a promising alternative to the use of organic media in enzymatic catalysis (*see* Chapters 24 and 25). Among the advantages associated with $SCCO_2$ are its nontoxicity, nonflammability, and relative cheapness. In addition, diffusivities in $SCCO_2$ are high, and the SC fluid offers the possibility of an integrated separation process *(1)*.

Basic requirements for a successful process are relevant solubilities of substrates and good stability of the enzyme preparation. The progress of the reaction is influenced by temperature, pressure, and water activity of the medium. Because one has to test several enzymes and substrates to achieve a high-reaction velocity, it is generally not possible to assess the influence of these parameters on each individual reaction. While the optimum temperature for a certain enzyme is generally known, the pressure should be chosen to ensure reasonably high solubility of the substrates. Optimum water activity will vary with the chosen enzyme and with the support in the case of immobilized enzyme preparations. If no information is available about how much water is required by the enzyme, we propose to perform two sets of experiments: one using dry CO_2 and one using CO_2 with a high water content close to saturation (*see* **Notes 1–4**).

Here we describe a method to achieve an enzymatically catalyzed transesterification, where we employ, as an example, the alcohol D,L-menthol and the ester isopropenyl acetate. Reaction products are L-menthyl acetate (the reaction proceeds enantioselectively) and acetone. Isopropenol isomerizes to give acetone and thus shifts the reaction equilibrium to the product side. The purpose of this specific reaction is to separate the racemic mixture of L- and D-menthol. This aspect is, however, not relevant for the method we are describing.

From: *Methods in Biotechnology, Vol. 13: Supercritical Fluid Methods and Protocols*
Edited by: J. R. Williams and A. A. Clifford © Humana Press Inc., Totowa, NJ

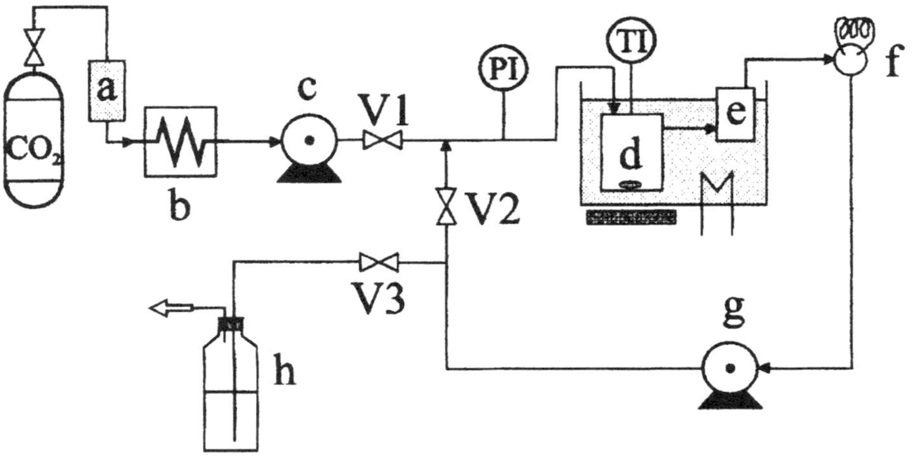

Fig. 1. Experimental system for enzymatic catalysis in supercritical carbon dioxide: *(a)* filter; *(b)* cryostat; *(c)* pressurizing pump; *(d)* reaction vessel with magnetic stirrer; *(e)* humidity sensor; *(f)* sampling and injection valve; *(g)* circulating pump; *(h)* liquid trap; P1 pressure transducer; T1 temperature transducer; V1, V2, and V3 valves.

2. Materials

1. CO_2, D,L-menthol, isopropenyl acetate, esterase EP10.
2. Reactor, pressurizing pump, magnetic gear circulating pump, thermostat, cryostat, valves, filter, sampling/injection valve (HPLC type) with 500 µL loop, magnetic stirrer.
3. 500-mL flask for use as a liquid trap.
4. Aluminum oxide humidity sensor, pressure, and temperature transducers.
5. A gas chromatograph, e.g., a Hewlett-Packard Series II.

3. Method

1. Assemble the enzymatic reaction system as shown in **Fig. 1**.
2. Prepare 200 mg of esterase EP10 (*see* **Note 5**).
3. Heat the water bath to 50°C and cool the cryostat to –10°C.
4. For 200 mg of enzyme preparation, the stirrer bar and D,L-menthol to a final concentration of 20 m*M* are introduced into the reactor and the reactor is connected to the system (*see* **Note 6**).
5. The system is flushed with several volumes of CO_2 with V1 and V3 open and V2 closed.
6. Close V3 when the pressure rises to 50 bar.
7. Start the piston pump and operate until the pressure reaches 100 bar. Then close V1 and open V2.
8. Start the magnetic stirrer at 300 rpm and the magnetic gear circulating pump.
9. Readjust the pressure to 100 bar, after constant values of pressure and temperature have been reached.

10. Add 1 mL of isopropenyl acetate, corresponding to a final concentration of 50 m*M*, via the HPLC-valve. This marks the start of the reaction.
11. Take samples after 15 min, and 2, 4, 6, 8, and 24 h. The content of the 500 μL sample loop is expanded into 2 mL of hexane in a 5-mL graduated flask (not shown in **Fig. 1**). The graduated flask is filled to the mark, the solution transferred to a screwcap flask and stored at 4°C.
12. Analyze the samples on the gas chromatograph (*see* **Note 7**).
13. After the last sample has been withdrawn, stop the magnetic gear circulation pump and the magnetic stirrer.
14. Depressurize the system slowly, by closing V2 and opening V3 carefully, bubbling the CO_2-stream through 300 mL of ethanol in the liquid trap. This will take several hours.
15. Clean the system by disconnecting the reactor and washing out with water and acetone. Then fill with ethanol and connect to the system. Pressurize the system from the CO_2 supply pump via V1, with V3 closed and V2 open. Adjust the pressure to 150–200 bar, start the magnetic stirrer and circulating pump for 1 h to recirculate the mixed ethanol and CO_2. Repeat this procedure with the reactor filled once again with ethanol and once more with acetone. Finally, disconnect the reactor and flush the piping with CO_2 for 30 min.

4. Notes

1. The solubilities of the substrates as a function of pressure and temperature should be known in advance. As a rule of thumb, the solubility of organic low molecular weight compounds containing oxygen is generally high. Nonpolar, low molecular weight compounds show moderate solubility, and polar organic compounds of high molecular weight are nearly insoluble.
2. Enzymes are generally more thermostable in supercritical media and in organic solvents than in aqueous solutions because of their high conformational rigidity in the absence of water *(2,3)*. Consequently, maximum activity is also shifted toward higher temperature. This effect is of course less pronounced as water activity rises.
3. A rise in pressure at constant temperature is always accompanied by a rise in density and, as a consequence, solubilities also increase. If substrates are easily soluble, we suggest that a moderate pressure (100 bar) is chosen because, in general, reaction velocities decrease at higher pressure (and therefore at constant substrate concentrations) *(4)*.
4. The influence of water activity can hardly be overestimated. If the enzyme contains too little water, it will be practically inactive, but also a water content that is too high will result in a considerable loss in activity *(5,6)*. We decided to measure the water vapor pressure (from which water activity is easily calculated) because water activity reflects the amount of water bound to the enzyme, which is the crucial parameter *(7)*. At the same water activity, hexane will have a much lower water content than the more polar $SCCO_2$. The amount of water bound to an enzyme suspended in the two solvents will, however, be the same.

Note that in a transesterification, no water is liberated or consumed. Water activity is thus not expected to change in the course of the reaction.

5. Esterase EP10 from *Pseudomonas marginata* was prepared at the institute of biotechnology at the Technical University of Graz *(8)*. The lyophilized powder was used as received.
6. In this example, no water is added. If the addition of water is necessary, it can be introduced together with the enzyme and substrates or it may be added via the HPLC-valve, when the system is already under pressure. To add a certain amount of water via this valve, it can be equipped with sample loops of different sizes. The solubility of water in $SCCO_2$ at a certain pressure and temperature can be calculated according to Chrastil *(9)*.
7. Another possibility is to use a supercritical fluid chromatograph for on-line measurements.

References

1. Aaltonen, O. and Rantakylä, M. (1991) Biocatalysis in supercritical carbon dioxide. *Chemtech* **21,** 240–248.
2. Kamat, S., Critchley, G., Beckmann, E. J., and Russell, A. J. (1995) Biocatalytic synthesis of acrylates in organic solvents and supercritical fluids: 3. Does carbon dioxide covalently modify enzymes? *Biotechnol. Bioeng.* **46,** 610–620.
3. Volkin, D. B., Staubli, A., Langer, R., and Klibanov, A. M. (1990) Enzyme thermoinactivation in anhydrous organic solvents. *Biotechnol. Bioeng. B* **37,** 843–853.
4. Kamat, S. V., Iwaskewycz, B., Beckmann, E. J., and Russell, A. J. (1993) Biocatalytic synthesis of arylates in supercritical fluids: tuning enzyme activity by changing pressure. *Proc. Natl. Acad. Sci. USA* **90,** 2940–2944.
5. de Carvalho, I. B., de Sampaio, T. C., and Barreiros, S. (1996) Solvent effects on the catalytic activity of subtilisin in compressed gases. *Biotechnol. Bioeng.* **49,** 399–404.
6. Marty, A., Chulalaksananukul, W., Willemot, R. M., and Condoret, J. S. (1992) Kinetics of lipase-catalyzed esterification in supercritical carbon dioxide. *Biotechnol. Bioeng.* **39,** 273–280.
7. Michor, H., Marr, R., and Gamse, T. (1996) Enzymatic catalysis in supercritical carbon dioxide: effect of water activity, in *High Pressure Chemical Engineering, Process Technology Proceedings 12* (Rudolf von Rohr Ph. and Trepp, Ch., eds.), Elsevier, Amsterdam, pp. 115–120.
8. Stubenrauch, G., Griengl, H., Klempier, N., Faber, K., and Schwab, H. (1995) Esterase aus *Pseudomonas marginata*. Patent no. AT-399.886.
9. Chrastil, J. (1982) Solubility of solids and liquids in supercritical gases. *J. Phys. Chem.* **86,** 3018–3021.

24

Transesterification Reactions Catalyzed by Subtilisin Carlsberg Suspended in Supercritical Carbon Dioxide and in Supercritical Ethane

Teresa Corrêa de Sampaio and Susana Barreiros

1. Introduction

The characteristics that make supercritical fluids attractive solvents for extraction also make them potentially interesting solvents for biocatalysis *(1)* (*see* Chapters 23 and 25). Of these, the possibility to integrate the steps of reaction and downstream separation is certainly one of the most appealing.

Carbon dioxide is the preferred supercritical fluid mainly because of its nontoxicity in a broad sense—e.g., CO_2 is allowed in the food industry—and its nonflammability. The latter characteristic is especially important for safety reasons when large amounts of solvent are needed in a process, as is often the case with extraction. For smaller-scale applications, the use of other solvents, such as ethane, could become an option. Indeed, CO_2 has been shown to have a negative effect on the catalytic activity of some lipases and proteases, either in free or immobilized form *(2–6)*. While there are authors who suggest that the adverse effect of CO_2 results from local pH changes on the enzymes *(3)*, others believe the main reason for impaired catalytic performance in this solvent is the formation of complexes between certain residues on the enzyme and CO_2 *(2,4)*.

The catalytic activity of many enzymes, including subtilisin Carlsberg, has been shown to depend strongly on their degree of hydration. Indeed, the activity/enzyme hydration profile for subtilisin in nonaqueous media is bell-shaped *(5,7–9)*; differences among solvents being more pronounced at the ends of the bell. Hence, it is very important to quantify enzyme hydration and to refer catalytic activity to a specific value of that parameter. One possibility,

From: *Methods in Biotechnology, Vol. 13: Supercritical Fluid Methods and Protocols*
Edited by: J. R. Williams and A. A. Clifford © Humana Press Inc., Totowa, NJ

described below, is to allow the enzyme to equilibrate with the solvent, measure the water content of the solvent after equilibration, combine this value with the water contents of the enzyme and the solvent before equilibration, and calculate the water content of the enzyme at equilibrium through a mass balance. More elaborate methods include the use of NMR ***(10)*** and of radiolabeled isotopes ***(11)***. Enzyme hydration may also be adjusted by varying the water activity in the medium through direct addition of pairs of salt hydrates ***(12)***.

CO_2 and ethane are essentially nonpolar solvents and thus are particularly adequate media for nonpolar species—e.g., CO_2 at 35°C and 100 bar has a solubility parameter of approximately 6.0 ***(13)***, as compared to 7.3 for *n*-hexane at ambient conditions ***(14)***. It is important to make sure that the concentrations of the substrates used have a safety margin relative to the solubility limit at the conditions of the experiments. Solubilities that are not found in the literature may be measured by following a procedure similar to that indicated for water (*see* **Subheading 4., step 8**).

The apparatus described below may be used at pressures up to 300 bar. It is important that there are no leaks, and hence the apparatus should be pressure tested with nitrogen after replacing parts. Of the safety precautions required, the use of safety glasses is of foremost importance. The protocol described here is a general method for transesterification reactions catalyzed by subtilisin Carlsberg and conducted in supercritical fluid media.

2. Materials

1. Subtilisin Carlsberg (from Bacillus licheniformis).
2. Carbon dioxide, ethane, and nitrogen with purities of ≥ 99.95 mol%.
3. Molecular sieves, 0.3-nm pore diameter.
4. Ice in a plastic bag (for cooling a pump-head).
5. Teflon bar (for pushing the Teflon piston out of the cell at the end of an experiment).
6. Lyophilization equipment.
7. Safety glasses whenever working at high pressure, gloves when manipulating organic solvents and substrates.
8. Experimental apparatus. **Fig. 1** shows a schematic of the experimental apparatus used for studies with enzymes in supercritical carbon dioxide. The CO_2 from the gas bottle goes through a line filter, is compressed and enters the ballast with molecular sieves as a liquid. The pressure in the CO_2 line is measured with a pressure transducer. Valve ADM is a chromatographic valve for admission of the substrates, using syringe S1. This valve is connected to the high-pressure cell via valve V6. Also connected to the cell is the chromatographic valve SAMP for taking samples. Valve V8 is a two-way valve that allows the release of samples directly into the titration chamber of a Karl-Fischer water titrator, and collection of samples in a flask for GC analysis. Valve SAMP is connected via valve V7 either to the syringe S2 containing an appropriate solvent for taking samples for

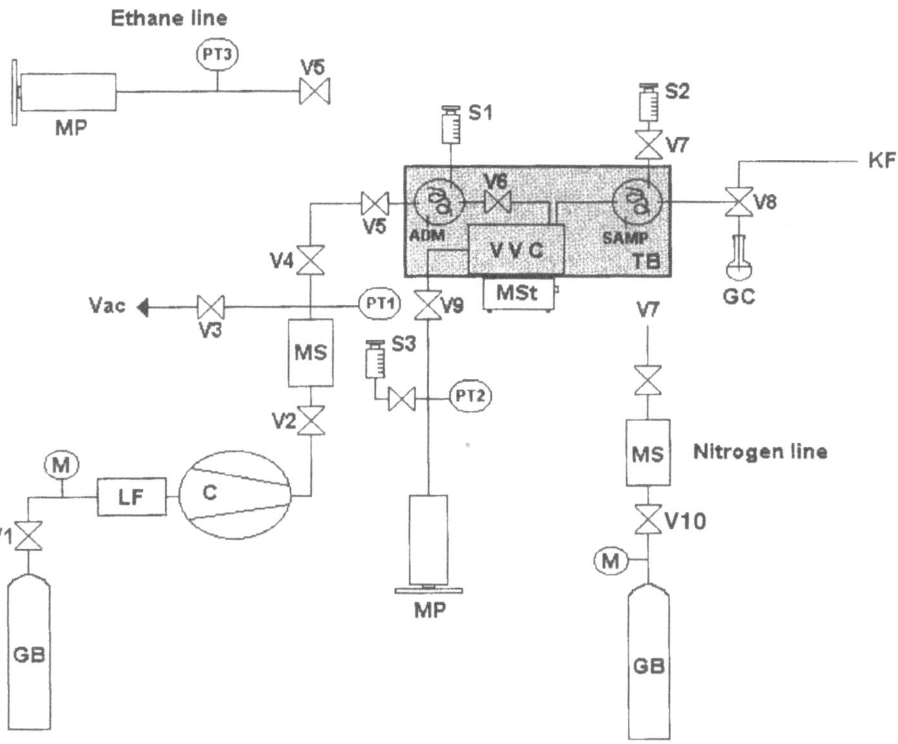

Fig. 1. Diagram of the experimental apparatus. GB, gas bottle; M, manometer; LF, line filter; C, compressor; MS, ballast with molecular sieves; PT1, PT2 and PT3, pressure transducers; Vac, vacuum line; ADM and SAMP, chromatographic switching valves; S1, S2 and S3, syringes; VVC, variable volume high pressure cell; MSt, magnetic stirrer; TB, thermostatted bath; KF, line for connection to the Karl-Fischer titrator; GC, line for taking samples for GC analysis; MP, manual syringe pump; V1-V10, needle valves.

GC analysis, or to a nitrogen line. Low-humidity nitrogen further dried over molecular sieves is used as a rinsing fluid when taking samples for Karl-Fischer analysis. The back-pressure fluid is pressurized with a manual syringe pump. The pressure of the reaction mixture is known indirectly via the pressure of this fluid, as indicated by pressure transducer PT2. The cell assembly comprises the high-pressure cell and all the parts connected to it, up to but excluding valves V5 and V9, which are permanently connected to the CO_2 and the back-pressure fluid lines, respectively. The cell, valve V6, and the loops of valves SAMP and ADM are immersed in a thermostatted liquid bath. When using ethane as solvent, the ethane line replaces the CO_2 line up to and including valve V5. The ethane is compressed with a manual syringe pump similar to that used for the back-pressure fluid (*see* **Note 1**).

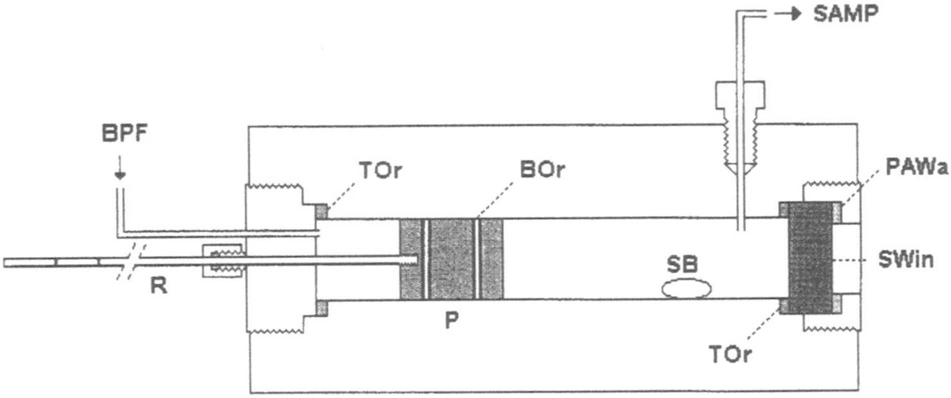

Fig. 2. Diagram of the variable-volume, high-pressure cell. BPF, back-pressure fluid; R, metal rod; TOr, Teflon O-rings; P, Teflon piston; BOr, Buna *M* O-rings; SB, Teflon-coated stir bar; PAWa, polyacetal washer; SWin, sapphire window; SAMP, line for connection to valve SAMP.

Fig. 2 shows a schematic of the variable-volume, high-pressure stainless steel cell. The seal at both ends is provided by a Teflon O-ring. The polyacetal washer avoids damage to the window. The tubing connected to the back-pressure fluid line is soldered to the rear end screw of the cell. The Teflon piston with Buna *M* O-rings separates the back-pressure fluid from the reaction mixture. A stainless steel rod with marks corresponding to well-defined volumes of the cell is screwed onto the rear end of the piston, going through a nut with Teflon ferrules to allow the movement of the rod. The connection to valve V6 is behind the plane of the drawing, at a 45° angle with the connection to valve SAMP.

Figs. 3 and **4** show schematically how the chromatographic valves SAMP and ADM are used. In position 1, the loop is filled with either a sample (valve SAMP) or a mixture of substrates (valve ADM); in position 2, the contents of the loop are discharged either for Karl-Fischer titration or GC analysis (valve SAMP), or are admitted into the cell.

3. Method

1. Take the cell assembly and the loose parts (sapphire window, polyacetal washer, piston, Teflon O-rings, Buna *M* O-rings) from the oven and allow them to cool down (*see* **step 15**).
2. Place the Buna *M* O-rings on the piston and push it into the cell, from the back, so that the rear end of the piston stays about 5 mm below the bed of the Teflon O-ring. Screw the metal rod onto the piston. Put the Teflon O-ring in place. Allow the rod to go through the small nut with Teflon O-rings and screw on the rear end screw of the cell, with the cell held tightly in a bench vice (take care not to bend the tubing for connection to the back-pressure fluid too close to the soldered

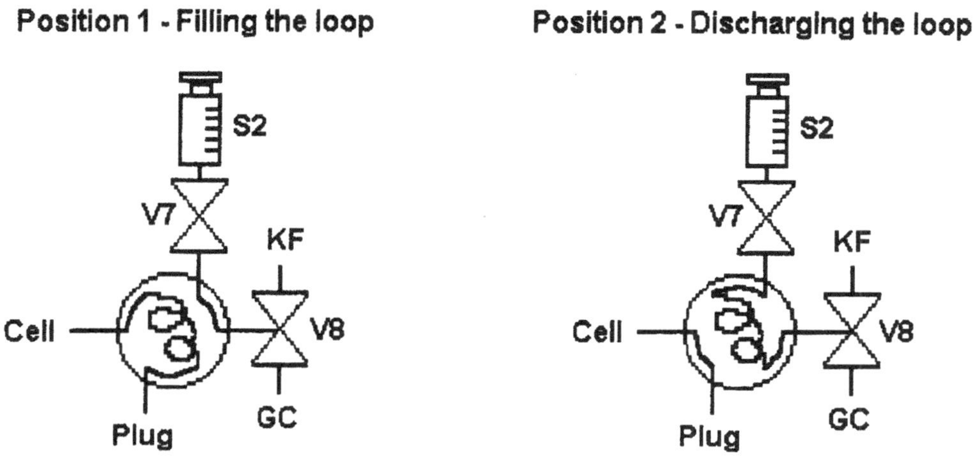

Fig. 3. Diagram of the chromatographic valve SAMP. Notation as in **Fig. 1**.

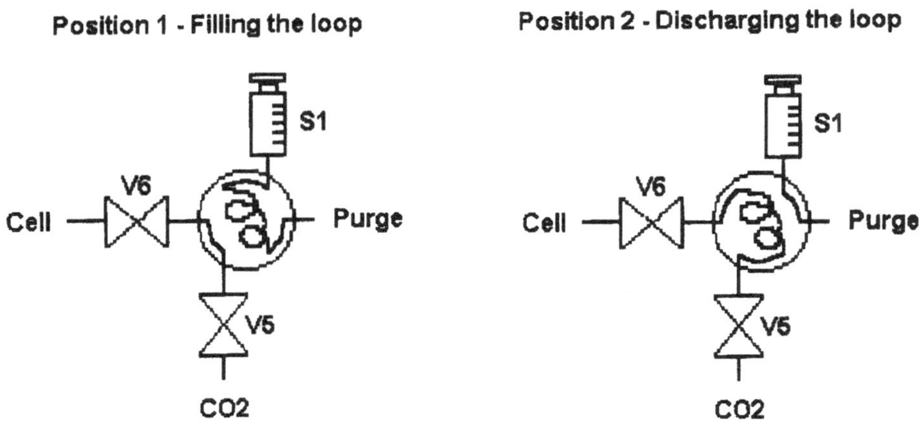

Fig. 4. Diagram of the chromatographic valve ADM. Notation as in **Fig. 1**.

joint). Adjust the torque on the small nut so as to avoid leaks of the back-pressure fluid, while allowing the rod to move with the piston.

3. Hold the cell in the vice so that its front end is slightly tilted upward. Introduce the enzyme, water if needed (*see* **Notes 2–4**), and a stir bar (a better enzyme suspension is obtained in the presence of the alcohol substrate). Put the Teflon O-ring in place. Put the polyacetal washer in position on the front end screw of the cell where it should have a tight fit and let the sapphire window rest on it. Screw on the front end screw. Do not connect syringe S2 to valve V7. Do not connect the KF line of valve V8 to the Karl-Fischer apparatus. Do not connect syringe S1 to valve ADM.

4. Place the cell assembly in a thermostatted bath. Connect the cell to the solvent admission line, via valve V5. Connect the back of the cell to the back-pressure fluid line, via valve V9.
5. Pressurize the back-pressure fluid with the manual pump. By gently opening valve V9, let the back-pressure fluid push the piston a little bit forward (follow the movement of the rod). Close valve V9. Open valves V1, V2, and V4 and pressurize the CO_2 to a pressure slightly higher than that of the back-pressure fluid, as indicated by pressure transducer PT2. Make sure valve ADM is in position 1 in **Fig. 4**. Open valve V6 gently, and admit a small amount of the liquified solvent into the cell. The piston will go backward somewhat (make sure the piston is never pushed back completely against the rear end screw of the cell). Close valve V6 and admit more back-pressure fluid via valve V9. Close valve V9. Add more solvent via valve V6. Repeat these two operations slowly until the cell is filled with solvent at the desired pressure and volume, as indicated by the pressure transducer PT2 and the marks on the metal rod. The final adjustments should be made once the temperature of the cell has stabilized, which takes place about 1 h after starting the admission of the solvent.
6. Close valve V6. Keep valve V9 open and monitor the desired pressure with the pressure transducer PT2 (make sure there are no leaks). Make sure valve SAMP is in position 2 in **Fig. 3**. Stir the reaction mixture for about 2 h.
7. Connect valve V7 to the nitrogen line. Adjust the pressure regulator of the nitrogen bottle to a relative pressure of 0.5 bar, open valve V7, and then the Karl-Fischer side of valve V8. Allow the nitrogen to flow through the lines and into the atmosphere for about 10 min to eliminate any humidity that may exist. With nitrogen still flowing, immerse the KF line in the solution of the Karl-Fischer apparatus and allow the nitrogen to bubble through the solution until a constant drift is reached (the drift is the amount of water, in micrograms, that enters the titration chamber per minute). Register the drift (the average drift in case it fluctuates slightly around a constant value)—drift 1. Stop stirring the reaction mixture. While the enzyme settles down, close valves V7 and V8 and wait until the drift reaches a constant value—drift 2.
8. To take a sample for Karl-Fischer titration, turn valve SAMP to position 1 (*see* **Notes 5** and **6**). Immediately compensate for the pressure drop by pressurizing the back-pressure fluid further. With pressure back to its initial value, turn valve SAMP to position 2. Slowly turn the appropriate handle of valve V8 so that the contents of the loop are discharged directly into the titration chamber of the Karl-Fischer titrator. Once the gas stops bubbling through the solution, open valve V7 and rinse the expansion zone with nitrogen. Register the time interval between the opening of valve V7 and the endpoint of the titration - Δt. Close valves V7 and V8. Wait 10 min and take a second sample for Karl-Fischer titration. Repeat for a third sample (*see* **Notes 7–10**).
9. To proceed for the reaction, mix the ester substrate with appropriate amounts of the alcohol substrate and water (*see* **Note 11**.). Prepare about three times as much of this mixture as the volume of the loop of valve ADM. Connect syringe S1 to

valve ADM. Introduce the mixture in syringe S1 and first wash and then fill up the loop with mixture. Make sure the pressure of CO_2, as indicated by the pressure transducer PT1, exceeds that of the back-pressure fluid. Turn valve ADM to position 2. Open valve V6 slowly, thereby washing the contents of the loop into the cell with solvent. Stirring should be on. This marks the start of the reaction. As before, alternately admit CO_2 and back-pressure fluid so as to reach the desired pressure and volume, at the selected temperature.

10. Disconnect syringe S1. Connect the S1 and purge lines of valve ADM to a hood (e.g. with plastic tubing) and turn valve ADM to position 1. Close valves V6, V5, V4, V2, and V1, while keeping valve V9 open. Resume stirring.
11. Periodically, stop stirring, wait until the enzyme settles down and take a sample to follow the reaction. To do this, connect syringe S2 to valve V7 and introduce an appropriate amount of the desired solvent in the syringe. Take a volumetric flask with some of the solvent and immerse the GC line in it. By slowly opening valve V8, allow the gas to bubble through the solvent (it is important that the gas be released very gently). Once the gas stops bubbling, open valve V7 and wash the expansion zone with the solvent in syringe S2, collecting it in the flask. Disconnect syringe S2, fill it up with air, reconnect it to the circuit and push the solvent still in the lines into the collection flask. Close valves V7 and V8. Typically, six samples are drawn for GC analysis (*see* **Note 12**).
12. At the end of an experiment, with valve V9 open, depressurize the back-pressure fluid down to zero pressure. The piston will move backward and a meniscus indicative of the vapor–liquid equilibrium of the solvent will form. Close valve V9. Disconnect the cell assembly from valves V5 and V9, and take it out of the thermostatted bath.
13. In a hood, open valve V6 slowly and release the solvent. Unscrew the nut with Teflon ferrules which holds the metal rod just so it will be possible to unscrew the rear end-screw of the cell. Remove this screw and the Teflon O-ring, which may or may not be reutilized depending on the pressure at which the experiment was performed (e.g., O-ring too deformed after being extruded at 300 bar and hence discarded). Unscrew the front end-screw of the cell, remove the Teflon O-ring (same considerations as before), the polyacetal washer and the sapphire window. Use the Teflon bar referred to in **Subheading 2.** to push the piston backward, out of the cell. Remove the Buna *M* O-rings carefully with a stylus and discard them.
14. Wash the interior of the cell, all the lines of the cell assembly, loops, piston, O-rings, washer, sapphire window, first with water and then with acetone, with the help of a syringe. Make sure no solids remain in the tubing. Blow out the remains of acetone, first with a syringe, then with nitrogen (compressed air usually has a higher level of humidity).
15. Place the cell assembly and the loose parts—sapphire window, polyacetal washer, piston, Teflon O-rings, Buna *M* O-rings—in an oven at 60°C, and leave them there to dry for about 8 h.
16. Make sure the liquid substrates are stored over molecular sieves.

4. Notes

1. To use ethane as solvent, first cool down the head of the appropriate manual syringe pump with ice. When the pump-head is cold (which should take about

half an hour), connect valve V5 of the ethane line to the ethane bottle which should be mounted upside down. Check for leaks. If there are none, slowly move back the piston of the pump so as to fill up the pump with liquid ethane (leave a safety margin; see below). The transfer should take place at the vapor pressure of ethane at the temperature of the pump-head. Close valve V5 of the ethane line and disconnect this valve from the ethane bottle. Move the piston back completely and remove the ice. As the temperature of the pump increases, so will the pressure. Make sure that the safety limits are always observed. When pressure (as indicated by the pressure transducer PT3) stabilizes, connect the cell assembly to valve V5.

2. The water concentration in the solvent mixture in the cell, at water partitioning equilibrium, $[H_2O_{eq}^{solv}]$, is given by

$$[H_2O_{eq}^{solv}] = [\text{KF reading} - \Delta t \times (\text{drift 1} - \text{drift 2})]/\text{volume of sampling loop.}$$

Here, $[H_2O_{eq}^{solv}]$ is in grams of water per cubic decimeter of solvent mixture, KF reading is in micrograms of water, Δt is in minutes, drift 1 and drift 2 are in micrograms of water per minute, and the volume of sampling loop is in microliters. The meaning of the parameters Δt, drift 1 and drift 2 is given in **Subheading 3., steps 7** and **8**.

3. The initial water concentration in the solvent, $[H_2O_{in}^{solv}]$, is monitored in separate experiments (filling the cell with just solvent and taking samples for Karl-Fischer titration).
4. The state of hydration of the enzyme at water partitioning equilibrium, $\%\ Hyd_{eq}^{enz}$, is given by a mass balance:

$$\%\ Hyd_{eq}^{enz} = [(\text{total water introduced in cell} - \text{water in solvent mixture at equilibrium})/\text{weight of dry enzyme}] \times 100$$

total water introduced in cell,

$$\text{in g} = (m_{in}^{enz} \times \%\ Hyd_{in}^{enz}/100) + [H_2O_{in}^{solv}] \times \text{volume of cell} + [H_2O^{alc}] \times \text{volume of alcohol}$$

Here, m_{in}^{enz} is the weight of lyophilized enzyme introduced into the cell, in grams, $\%\ Hyd_{in}^{enz}$ is the initial hydration of the lyophilized enzyme, as determined by direct Karl-Fischer titration of the powder, and $[H_2O^{alc}]$ is the water concentration in the alcohol substrate, also determined by direct Karl-Fischer titration. The volumes are in cubic decimeters.

$$\text{water in solvent mixture at equilibrium} = [H_2O_{eq}^{solv}] \times \text{volume of cell.}$$

$$\text{weight of dry enzyme} = m_{in}^{enz} - (m_{in}^{enz} \times \%\ Hyd_{in}^{enz}/100).$$

5. Samples for Karl-Fischer titration should be taken once the water partitioning equilibrium between enzyme and solvent mixture has been established. Before settling for a 2-h wait for this purpose, longer equilibration times were tested which led to the same results.
6. Taking a sample for Karl-Fischer titration involves the careful opening of valve V8 to release the gaseous solvent into the titration chamber of the apparatus,

waiting until the gas stops bubbling *(a)*, opening valve V7 and rinsing with nitrogen *(b)*. Between *(a)* and *(b)*, there is a moment when the water intake is low. In order to avoid this moment being considered as the end point of the titration, an adequate value for the delay function of the Karl-Fischer apparatus should be selected.

7. Sometimes, the first of the three samples taken for Karl-Fischer titration gives a value that is too high and has to be discarded. This most likely reflects insufficient drying of the sampling loop and accessing small diameter tubing.
8. The accuracy of the water quantification method should be checked periodically in separate experiments. To do this, known amounts of water are placed in the cell, which is then filled with solvent at a given temperature and pressure, and allowed to reach equilibrium. The water added initially is compared with the corresponding Karl-Fischer titration readings. It is important that both low and higher water contents of the solvent be probed. This is best done with CO_2 rather than with ethane, given the higher water solubility in the former solvent. In the moderate water concentration range, a simple procedure is to measure the solubility of water in CO_2 by placing in the cell an excess water phase, sampling the CO_2-rich upper phase at equilibrium and comparing with literature values.
9. The accuracy of the substrate and product quantification method should be similarly checked, by placing known amounts of the compounds in the cell, taking samples and comparing results.
10. The volume of the sampling loop should be kept as small as accurate sampling will permit, in order to avoid large pressure drops upon taking samples and possible momentary changes in the composition of the solvent mixture.
11. The second substrate is washed into the cell with solvent after the water content of the solvent mixture in the cell has been quantified by Karl-Fischer titration. To ensure that this addition does not change the concentration of the first substrate nor the water content previously determined, the second substrate should be mixed with appropriate amounts of the first substrate and water. The cell volume and the desired concentrations of both substrates and water allow the calculation of the total amount of each component that should be in that cell volume. Knowing how much of each component will have exited the cell upon taking the samples for Karl-Fischer titration (this depends on the number of samples taken and on the volume of the sampling loop), it is possible to calculate the amounts that should be added with valve ADM and dimension the loop of valve ADM.
12. The duration of an experiment aimed at the determination of an initial rate depends on the solvent, among other factors. Because subtilisin Carlsberg is much more active in ethane than in CO_2 at otherwise identical conditions, samples have to be taken at larger time intervals in the latter solvent. Experiments thus take longer in CO_2.

References

1. Vermüe, M. H. and Tamper, J. (1995) Biocatalysis in non-conventional media: medium engineering aspects. *Pure Appl. Chem.* **67,** 345–373.

2. Kamat, S., Barrera, J., Beckman, E. J., and Russell, A. J. (1992) Biocatalytic synthesis of acrylates in organic solvents and supercritical solvents: I. Optimization of enzyme environment. *Biotech. Bioeng.* **40,** 158–166.
3. Marty, A., Chulalaksananukul, W., Willemot, R. M., and Condoret, J. S. (1992) Kinetics of lipase catalyzed esterification in supercritical CO_2. *Biotech. Bioeng.* **39,** 273–280.
4. Kamat, S., Critchley, G., Beckman, E. J., and Russell, A. J. (1995). Biocatalytic synthesis of acrylates in organic solvents and supercritical fluids: III. Does carbon dioxide covalently modify enzymes? *Biotechnol. Bioeng.* **46,** 610–620.
5. Borges de Carvalho, I., Corrêa de Sampaio, T., and Barreiros, S. (1996) Solvent effects on the catalytic activity of subtilisin suspended in compressed gases. *Biotechnol. Bioeng.* **49,** 399–404.
6. Almeida, M. C., Ruivo, R., Maia, C., Freire, L., Corrêa de Sampaio, T., Barreiros, S., and Novozym, S. (1998) 435 activity in compressed gases: water activity and temperature effects. *Enzyme Microb. Technol.* **22,** 494–499.
7. Affleck, R., Xu, Z. F., Suzawa, V., Focht, K., Clark, D. S., and Dordick, J. S. (1992) Enzymatic catalysis and dynamics in low-water environments. *Proc. Natl. Acad. Sci. USA* **89,** 1100–1104.
8. Corrêa de Sampaio, T., Melo, R. B., Moura, T. F., Michel, S., and Barreiros, S. (1996) Solvent effects on the catalytic activity of subtilisin suspended in organic solvents. *Biotechnol. Bioeng.* **50,** 257–264.
9. Fontes, N., Nogueiro, E., Elvas, A. M., Corrêa de Sampaio, T., and Barreiros, S. (1998) Effect of pressure on the catalytic activity of subtilisin Carlsberg suspended in compressed gases. *Biochim. Biophys. Acta* **1383,** 165–174.
10. Parker, M. C., Moore, B. D., and Blacker, A. J. (1995) Measuring enzyme hydration in nonpolar organic solvents using NMR. *Biotechnol. Bioeng.* **46,** 452–458.
11. Gorman, L. A. S. and Dordick, J. S. (1992) Organic solvents strip water off enzymes. *Biotechnol. Bioeng.* **39,** 392–397.
12. Zacharis, E., Omar, I. C., Partridge, J., Robb, D. A., and Halling, P. J. (1997) *Biotechnol. Bioeng.* **55,** 367–374.
13. Hawthorne, S. B. (1990) Analytical-scale supercritical fluid extraction. *Anal. Chem.* **62,** 633 A-642 A.
14. Riddick, J. A., Bunger, W. B., and Sakano, T. K. (1986) Organic solvents: Physical Properties and Methods of Purification, 4th ed. Wiley-Interscience, New York.

25

Enzymatic Synthesis of Peptide in Water-Miscible Organic Solvent/Supercritical Carbon Dioxide

Hidetaka Noritomi

1. Introduction

Enzymatic catalysis in nonaqueous media (*see* Chapters 23 and 24) has revealed some beneficial features of enzymes such as enhanced thermostability and altered specificity, and thermodynamic equilibria are shifted to favor synthesis over hydrolysis, e.g., esterification and peptide formation *(1–4)*. Enzymatic reactions in water-miscible organic solvents have the advantage of the solubility of a variety of substrates, including amino acid derivatives, which are often poorly soluble in nonpolar solvents *(5)*. However, as the addition of a certain amount of exogenous water into a water-miscible organic solvent is required to obtain the enzyme activity, the yield of product at equilibrium is expected to be less than that in dry solvents, and, moreover, the stability of the enzyme is reduced by the autolysis. On the other hand, as the enzyme is insoluble in a nonaqueous medium, and is suspended, the enzymatic reaction tends to be a diffusion-controlled reaction *(4,6)*.

In this chapter, we present a protocol for improving water content and the diffusion-controlled process by adding supercritical carbon dioxide ($SCCO_2$) into acetonitrile containing small amounts of water, where α-chymotrypsin-catalyzed peptide synthesis between *N*-acetyl-L-tyrosine ethyl ester and glycinamide is carried out. The critical temperature and pressure of $SCCO_2$ are 31.3°C and 7.38 MPa, respectively, and the physical properties of $SCCO_2$, such as density, can be controlled by the temperature and the pressure of the system. $SCCO_2$ is nontoxic and has low viscosity and high diffusivity compared to conventional liquid solvents, and is used especially for food and pharmaceutical applications *(7)*.

From: *Methods in Biotechnology, Vol. 13: Supercritical Fluid Methods and Protocols*
Edited by: J. R. Williams and A. A. Clifford

2. Materials

1. Enzyme catalyst: bovine pancreatic α-chymotrypsin (CT) (EC 3.4.21.1) (Sigma, St. Louis, MO) (*see* **Note 1**).
2. Substrate solution consisting of 2.8 mg/mL *N*-acetyl-L-tyrosine ethyl ester (Ac-Tyr-OEt) (Sigma), 2.35 mg/mL glycinamide hydrochloride ($Gly\text{-}NH_2{\cdot}HCl$) (Sigma), and 3 µL/mL triethylamine (Et_3N) in acetonitrile (*see* **Note 2**).
3. A reactor for high-pressure biocatalytic reactions (e.g., a supercritical fluid extraction/chromatography model super-200 system-3, Jasco) (*see* **Note 3**), with a magnetic stirrer.
4. A high-pressure cell of 50 mL internal volume (*see* **Note 3**).
5. CO_2 (purity exceeding 99.9%) (*see* **Note 4**).
6. High-performance liquid chromatograph with UV detector.
7. C18 column for high-performance liquid chromatography (HPLC) (e.g., a Capcell-Pak C18, 4.6 mm × 150 mm, Shiseido).
8. Mobile phase for HPLC: water/acetonitrile (4:1 by vol.) (*see* **Note 5**).
9. Membrane filter: polytetrafluoroethylene membrane filter (pore size 0.65 µm) (e.g., from Millipore).

3. Method

1. Place 40 mg of CT, 2.5 mL of distilled water, 37.5 mL of substrate solution, and a magnetic Teflon-coated bar in the high-pressure cell.
2. Flush CO_2 into the high-pressure cell (*see* **Note 6**). Pump CO_2 into the high-pressure cell with a high-pressure pump until the pressure in the system is 20 MPa (*see* **Note 7**).
3. After the high-pressure cell is filled with $SCCO_2$, stir the reaction mixture with the magnetic bar vigorously. Incubate the reaction mixture at 35°C and 20 MPa for 5 h.
4. After 5 h of incubation, stop stirring the reaction mixture with a magnetic bar. Depressurize the loaded $SCCO_2$ to atmospheric pressure by controlling the back-pressure regulator. Bubble the gas through acetonitrile in the collector. After the pressure in the system reaches atmospheric pressure, withdraw the reaction mixture remaining in the high-pressure cell and acetonitrile in the collector.
5. Filter the reaction mixture with the polytetrafluoroethylene membrane filter. Dilute the filtrate with the mobile phase (HPLC), and inject an aliquot of the solution into the HPLC instrument (*see* **Note 8**).

4. Notes

1. The enzyme should be stored at –20°C, and used as soon as possible after purchase because it is a protease, and the autolysis easily occurs.
2. Ac-Tyr-OEt and $Gly\text{-}NH_2{\cdot}HCl$ should be stored in a refrigerator. The substrate solution should be prepared just before use. Although Et_3N is not a substrate, it should be added to the reaction mixture in order to eliminate hydrochloric acid on the amino group in $Gly\text{-}NH_2{\cdot}HCl$, since, in a peptide formation reaction, the true nucleophile is considered to be the amino acid amide with a free amino group *(8)*.

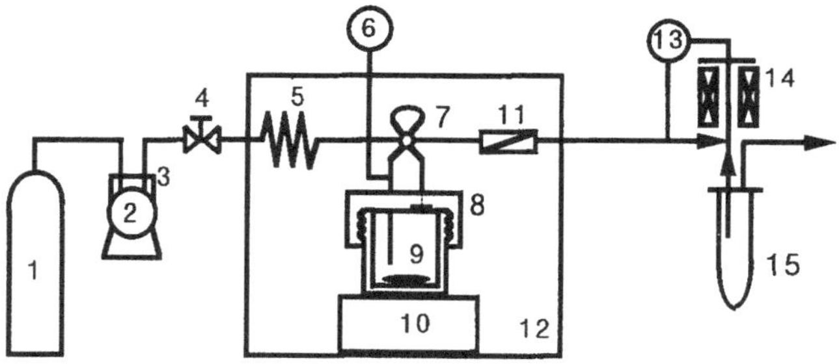

Fig. 1. Reactor configuration for high-pressure biocatalytic reactions: *1,* CO_2 cylinder; *2,* high-pressure pump; *3,* cooling jacket; *4,* valve; *5,* preheating coil; *6,* pressure gauge; *7,* 6-way valve; *8,* high-pressure cell; *9,* magnetic bar; *10,* magnetic stirrer; *11,* filter; *12,* air bath; *13,* pressure controller; *14,* back-pressure regulator; *15,* collector.

3. The reactor configuration is shown in **Fig. 1**. Liquid CO_2 is charged into a high-pressure HPLC pump and compressed to a desired pressure. The head of the pump is cooled by a cooling jacket in order to prevent the vaporization of liquid CO_2. Liquid CO_2 passes through preheating coil tubing in an air bath to reach the supercritical fluid condition. Supercritical CO_2 then enters a high-pressure cell. The pressure in the system is controlled by a back-pressure regulator having an accuracy of ±0.1 MPa, and monitored by a digital pressure gauge. The high-pressure cell is made of stainless steel.
4. The pressure of the CO_2 cylinder should be above 5 MPa.
5. The mobile phase should be degassed.
6. Make sure that the system does not leak before the experiment.
7. CO_2 from a cylinder is cooled at –7°C by a cooling jacket, compressed by a high-pressure pump, and then heated through a preheating coil tubing in an air bath whose temperature is 35°C. The pressure is controlled at 20 MPa by a back-pressure regulator.
8. The flow rate of mobile phase is 1.0 mL/min. The wavelength of the UV detector is 270 nm.

References

1. Klibanov, A. M. (1990) Asymmetric transformations catalyzed by enzymes in organic solvents. *Acc. Chem. Res.* **23,** 114–120.
2. Gupta, M. N. (1992) Enzyme function in organic solvents. *Eur. J. Biochem.* **203,** 25–32.
3. Noritomi, H., Almarsson, O., Barletta, G. L., and Klibanov, A. M. (1996) Influence of the mode of enzyme preparation on enzymatic enantioselectivity in organic solvents and its temperature dependence. *Biotechnol. Bioeng.* **51,** 95–99.
4. Wescott, C. R., Noritomi, H., and Klibanov, A. M. (1996) Rational control of enzymatic enantioselectivity through solvation thermodynamics. *J. Am. Chem. Soc.* **118,** 10,365–10,370.

5. Kise, H., Hayakawa, A., and Noritomi, H. (1990) Protease-catalyzed synthetic reactions and immobilization-activation of the enzymes in hydrophilic organic solvents. *J. Biotechnol.* **14,** 239–254.
6. Schmitke, J. L., Wescott, C. R., and Klibanov, A. M. (1996) The mechanistic dissection of the plunge in enzymatic activity upon transition from water to anhydrous solvents. *J. Am. Chem. Soc.* **118,** 3360–3365.
7. Hutchenson, K. W. and Foster, N. R. (1995) Innovations in supercritical fluid science and technology, in *Innovations in Supercritical Fluids* (Hutchenson, K. W. and Foster, N. R., eds.), American Chemical Society, Washington, D.C., pp. 1–31.
8. Kise, H., Fujimoto, K., and Noritomi, H. (1988) Enzymatic reactions in aqueous-organic media. VI. Peptide synthesis by α-chymotrypsin in hydrophilic organic solvents. *J. Biotechnol.* **8,** 279–290.

26

Micronization of a Polysaccharide by a Supercritical Antisolvent Technique

Alberto Bertucco and Paolo Pallado

1. Introduction

Recently, supercritical fluids (SFs) have been used in applications closely related to biotechnology. For example, they have been proposed as media for processing biocompatible polymers to develop new products for medical and pharmaceutical applications *(1)*. It was shown that SFs are suitable for the micronization of pharmaceuticals to obtain controlled drug-delivery systems characterized by particles with a desired small size, a narrow size distribution, and uniformly impregnated by the drug *(2)*.

The production of fine particles can be achieved by different SF techniques, such as the rapid expansion of a supercritical solution, RESS *(3)* (*see* Chapter 27), the particles from gas-saturated solution process, PGSS *(4)* and the supercritical antisolvent precipitation, SAS or GAS *(5)*. In general, CO_2 is the SF most widely considered and investigated, for its well-known favorable properties, but other compounds can be profitably used as well.

From the knowledge developed so far, it is also clear that the RESS technique is no more than a potential, due to the extremely low solubility of the compounds of interest in supercritical SFs, even at pressures as high as 500 bar. On the other hand, the PGSS process often requires temperatures too high for the stability of the drug. On the basis of different authors' and our own experiences, we believe that the SAS technique, which can be performed at pressures usually lower than 100 bar, is the only one likely to become of practical interest as an alternative to traditional technologies currently used for the production of fine biocompatible particles.

The possibility of exploiting dense CO_2 as an antisolvent was first conceived and devised for the fine comminution of high explosives by Gallagher et al. in

From: *Methods in Biotechnology, Vol. 13: Supercritical Fluid Methods and Protocols*
Edited by: J. R. Williams and A. A. Clifford © Humana Press Inc., Totowa, NJ

1989 *(5)*, but soon it was extended with success to other different fields; and a number of papers have been published about the use of SAS with biopolymers *(6–9)*. The SAS technique takes advantage of the SF antisolvent property, rather than the solvent capacity, which is often too low for practical purposes. For example, CO_2 does not exert any detectable solubility for polysaccharides of large molecular weight regardless of the pressure and temperature conditions used, but it can be solubilized to any extent in many organic solvents. The amount of CO_2 dissolved in the liquid is an increasing function of pressure, so that the properties of the organic solvents can be strongly modified by CO_2 addition up to a point where the mixture is no longer able to keep the polysaccharide in solution. At this point, a complete precipitation of the polysaccharide can be obtained by a further, but small, increase of pressure. The swelling caused by the action of CO_2 on most common organic solvents can be measured as the volumetric expansion of the liquid phase, and is a strong function of pressure and temperature. The behavior outlined above (i.e., expansion and subsequent precipitation by the action of an SF) is generally observed for most organic solvents, and many SFs are able to trigger this phenomenon. The knowledge of the expansion curve versus pressure and of the precipitation pressure at the selected temperature is essential for performing any supercritical antisolvent precipitation experiment. Both of them depend on the SF, organic solvent, solute, and temperature. On the other hand, the concentration of solute in the starting organic solution affects the shape and dimensions of the precipitate obtained.

This chapter describes how to carry out a batch SAS experiment, when the optimum conditions (i.e., pressure, temperature, and starting solution concentration) are known for the system of interest. The protocol describes the simplest way to determine the optimum conditions by performing the SAS experiment in a windowed cell. Example results are given. The general SAS procedure outlined below, with reference to the material described in **step 7** of the following section, has been patented *(10)*.

2. Materials

1. An apparatus for SAS. A simplified scheme of the SAS experimental apparatus is shown in **Fig. 1**. The main units are: a CO_2 reservoir; a chiller; a high-pressure HPLC-type pump; a high-pressure precipitation vessel (*see* **Note 1**); two needle valves to regulate the precipitation vessel pressure; a thermostatic system for temperature control in the precipitation vessel; a postexpansion vessel, operating at atmospheric conditions, to collect the organic solvent; a rotameter to check CO_2 flow rate, and a dry or humid test meter, to measure CO_2 total volume at standard conditions; high-pressure stainless steel tubing, fittings, and on/off valves, standard outer diameter 1/4 inch.

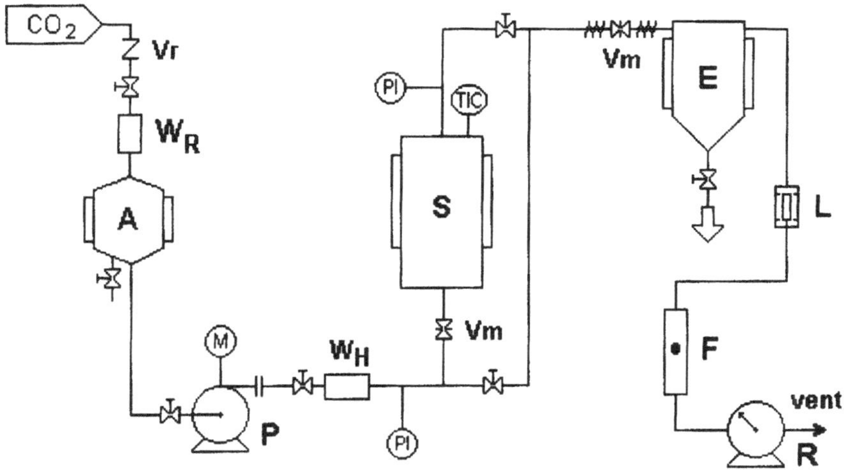

Fig. 1. Schematic diagram of the experimental system: *A*, CO_2 reservoir; *E*, expansion vessel; *F*, flow indicator; *L*, filter; *P*, CO_2 pump; *PI*, pressure indicators; *R*, flow meter; *S*, precipitation vessel; *TIC*, temperature controller; *Vr*, one-way valve; W_H, heat exchanger; W_R, chiller; and V_m, metering valve.

2. A filtration and distribution system. The inlet and outlet of the precipitator are equipped with stainless steel filters (frits) as distribution and filtration devices (*see* **Note 2**).
3. Safety equipment and devices. Pressure release and pressure safety systems are directly connected at the top flange (*see* **Note 3**).
4. A vent line for gas evacuation from the room.
5. A temperature sensor, which is sunk in the vessel through a 1/8 inch end connector, to get a more accurate regulation of the vessel internal temperature (*see* **Note 1**).
6. Manometers and/or pressure transducers (PI in **Fig. 1**) are used at various points on the apparatus. We suggest measuring pressure in at least two places: up and downstream of the precipitation vessel, to check possible blockage of the frits.
7. The biocompatible polymer stock solution. The solid polysaccharide (HYAFF, U.S. pat. 4,851,521, i.e., hyaluronic acid ethyl ester where all the acid carboxylic groups are esterified with the alcohol) is dissolved in dimethyl sulfoxide (DMSO), reagent grade (*see* **Note 4**). DMSO shows high solvent capacity toward HYAFF (250–270 mg/mL at 25°C) and high volumetric expansion when contacted with carbon dioxide.
8. Pure liquid CO_2 (purity>99.9%). The CO_2 has to be cooled and stored in an insulated vessel provided with an internal cooler.

3. Method

1. Prepare the polysaccharide liquid solution. The time required for dissolution of the polysaccharide in the organic solvent depends on the amount of polymer to

be added, but it requires a period of conditioning under stirring and heating (*see* **Note 4**).

2. Condition the supercritical apparatus by driving the chiller, thermostatic bath and heat exchanger to steady-state conditions, so that the desired temperature in the precipitator vessel is ensured before the run is started.
3. Load the solution prepared as in **step 1** into the precipitation vessel, which is then closed.
4. Pressurize the cell with CO_2. With the top line to the precipitation vessel (**Fig. 1**) closed, CO_2 is fed to the vessel from the bottom by opening the appropriate valves, closing the valve to the expansion vessel (**Fig. 1**), and activating the pump. CO_2 bubbles through the liquid solution with the frit acting as a distributor; the step is prolonged until the desired pressure (usually less than 100 bar) is reached (*see* **Note 5**). At this point, the precipitation of the polymer is complete. Note that a constant temperature is essential for the reproducibility of the experiment (*see* **Note 6**).
5. Wash the liquid phase (organic + CO_2 mixed solvent) out of the precipitation vessel (**Fig. 1**). This step is carried out at constant (maximum) pressure. The inlet of CO_2 is maintained from the bottom, but the top line is now opened to the expansion vessel (**Fig. 1**). The drainage of the liquid is thus obtained, while solid particles previously precipitated are trapped by the frit located at the top. The vessel pressure is regulated by an outlet valve to the expansion vessel (*see* **Note 7**).
6. Purify the product, removing the organic solvent absorbed/adsorbed on the particles by continuing to pass supercritical CO_2 through the precipitation vessel, S. The supercritical CO_2 is now being used as a solvent for the organic solvent. The time required depends on the precipitator temperature, the amount of starting solution, the SF flow rate, and the properties of the liquid solvent (*see* **Note 8**).
7. Depressurize the vessel by cutting off the supply of CO_2 and releasing the pressure through the top line and expansion vessel (*see* **Note 9**).
8. Collect the precipitated solid particle products, when atmospheric pressure is attained, by opening the vessel. The DMSO can be collected from the postexpansion chamber (*see* **Notes 10** and **11**).

4. Notes

1. The precipitator is a pressure vessel made by a main body (forged tube) and screwed flanges, all built in stainless steel. Its main features are: internal volume of about 0.2 dm^3; neoprene O-ring seals between vessel and flanges; PTFE gaskets for end connectors and internal metallic filters (frits); 1/4 inch tube connections of Swagelok® type; thermo-resistance sensor (Pt 100 Ω sunk in the vessel, connected through the top flange with a 1/8 inch end connector, Swagelok® type); thermal regulation of the vessel, obtained using an external jacket provided with inlet and outlet connectors for 6- to 10-mm rubber pipe, welded or screwed to the body. Note that the apparatus of **Fig. 1** can be modified to measure both the volumetric expansion and the incipient precipitation pressure by replacing the precipitation vessel with a windowed cell, for which a standard level

indicator, such as those used in high-pressure steam lines (i.e., Klinger cell type) can be easily adapted.

2. Frits with an average cutoff size lower than 3 μm are housed in both the top and bottom flanges and fixed by a screwed metal ring. A PTFE O-ring located between the frit and the flange body ensures a uniform and fine distribution of the inlet flux, avoiding the formation of preferential paths and leakage. The liquid is not able to seep through by gravity alone.
3. A manual valve is needed for fast draining and a rupture disk (or a check valve) prevents the occurrence of overpressure in the vessel. Both of them are inserted in the top flange.
4. It is necessary to maintain the solution under stirring and heating for a convenient time (more than 24 h for HYAFF in DMSO), depending on the amount of polymer used. The polymer has to be gradually added to the solvent to avoid agglomeration which increases the time required for dissolution. In our case, temperature was kept at 40°C and polymer concentration ranged from 0.2% to 1.5% by weight.
5. Pressure gradients from 5 to 20 bar/min were used during the pressurization step. The increase in pressure from ambient up to around 50 bar was controlled by the valve (Hoke, model 1325G4Y) between the CO_2 cylinder and the CO_2 reservoir. An HPLC pump (Rainin, model Dynamax SD 200) with a cooled head was then turned on, in order to further increase the pressure up to the desired value. For experiments with the windowed cell, it is necessary to operate with a lower pressure gradient, so that to perform a quasi-equilibrium process (remember that these runs are aimed to measure the volumetric expansion of the solution and the solute precipitation pressure).
6. During the run, the precipitation vessel temperature has to be regulated around the desired value to avoid overshootings due to both isoentropic compression and exothermic mixing of CO_2 in the liquid solution. This temperature should be not far from the critical temperature of the solvent; in our experiments with HYAFF, temperatures between 288.15 and 323.15 K were used.
7. Pressure was controlled manually by the valve (Hoke, model 1325G4Y), located on the entrance to the expansion vessel, E. Outlet CO_2 flow rates varied from 0.2 Nm^3/h up to 1.0 Nm^3/h. During this step, liquid DMSO was collected in the postexpansion vessel E, operating at atmospheric conditions, while CO_2 was vented to the flow metering section, which includes a rotameter (UCAR, model Matheson, type 7640T; measuring tube model 603) and a dry test meter (SMG Samgas Milano, type R/1).
8. Outlet CO_2 flow rate was controlled manually with the same valve as in previous **item 7**, with flow rates in the same range. The duration of this step can be extended in order to lower the residual content of DMSO in the final product down to the desired value.
9. Depressurization is preferably carried out by using pressure gradients ranging from 2–5 bar/min. Extreme care must be taken to prevent clogging of the needle valve before the expansion vessel, because of solids produced by the expansion effect.

Fig. 2. SEM photograph (bar = 1 µm, ×10,000) of HYAFF microspheres formed at the following conditions: initial concentration, 1.2% w/w HYAFF in DMSO; precipitator temperature, 313 K; final precipitation pressure, 90 bar; CO_2 flow rate, 8.0 g/min.

10. The yield of the process is very close to 100%: the mass of HYAFF microspheres collected was always equal to the amount initially loaded as a DMSO solution. The fraction of organic solvent recovered depends on the solvent volatility at the temperature of the postexpansion vessel; with DMSO, it was nearly 100%.
11. We discuss the results obtained by applying the SAS technique to the production of HYAFF micronic particles elsewhere *(2)*; we briefly summarize here the main achievements of the technique in this case. For this we refer to the product obtained and to the improvement gained in comparison with the current technology, which involves a solvent emulsion precipitation method (SEP). From **Fig. 2** it is clear that SAS provides HYAFF spherical particles with an average size of 400–500 nm, which is 50 times lower than the one obtained by SEP. In **Fig. 3**, the size distribution of these particles, as measured by a Coultard (for SAS) or Malvern (for SEP) analyzer, is presented and compared with particles produced by SEP: again, the improvement is remarkable. In addition, we note that SEP is a process that requires many more operating steps than SAS and that also the amount of organic solvents needed is dramatically reduced by SAS. Finally, we recall that SAS can be carried out in a semicontinuous mode, which allows precipitation at temperature and pressure steady-state conditions. On the other hand, this powerful and promising technique needs high-pressure apparatus and operation, which may represent a serious difficulty, especially for researchers not use to it.

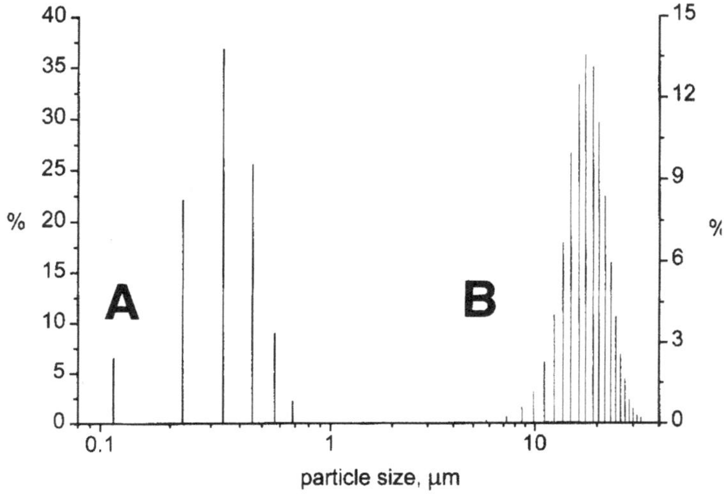

Fig. 3. Comparison of size distributions of particles obtained with **(A)** the SAS process and **(B)** SEP technique.

References

1. Debenedetti, P. G., Tom, J. W., Yeo, S. D., and Lim, G. B. (1993) Application of supercritical fluids for the production of sustained delivery devices. *J. Controlled Release* **24,** 27–44.
2. Benedetti, L., Bertucco, A., and Pallado, P. (1997) Production of micronic particles of biocompatible polymer using supercritical carbon dioxide. *Biotech. Bioeng.* **53,** 232–237.
3. Matson, D. W., Fulton, J. L., Petersen, R. C., and Smith, R. D. (1984) Rapid expansion of supercritical fluid solutions: solute formation of powders, thin films, and fibers. *Ind. Eng. Chem. Res.* **26,** 2298–2306.
4. Weidner, E., Steiner, R., and Knez, Z. (1996) Powder generation from polyethyleneglycols with compressible fluids, in *High Pressure Chemical Engineering* (von Rohr, R. and Trepp, C., eds.), Elsevier, Amsterdam, pp. 223–228.
5. Gallagher, P. M., Coffey, M. P., Krukonis, V. J., and Klasutis, N. (1989) Gas antisolvent recrystallization: new process to recrystallize compounds insoluble in supercritical fluids, in *Supercritical Fluid Science and Technology ACS Symp. Ser., No. 406* (Johnston, K. P. and Penninger, J. M. L., eds.), American Chemical Society, Washington, D.C., pp. 334–354.
6. Randolph, T. W., Randolph, A. D., Mebes, M., and Yeung, S. (1993) Submicrometer-sized biodegradable particles of poly(L-lactic acid) via the gas antisolvent spray precipitation process. *Biotech. Prog.* **9,** 429–435.
7. Yeo, S. D., Lim, G. B., Debenedetti, P. G., and Bernstein, H. (1993) Formation of microparticulate powders using a supercritical fluid antisolvent. *Biotech. Bioeng.* **41,** 341–346.

8. Falk, R. and Randolph, T. (1997) Controlled release of ionic pharmaceutical from poly (λ-lactide): microspheres produced by precipitation with a compressed antisolvent. *J. Controlled Release* **44,** 77–85.
9. Tomasko, D. L. and Chou, Y. H. (1997) Gas crystallization of polymer-pharmaceutical composite particles. *Proceedings of the Fourth International Symposium on Supercritical Fluids* I.S.A.S.F., Sendai (Japan), pp. 55–57.
10. *European Pat.* PCT/EP96/01354 (1996) Microspheres comprising a biocompatible polysaccharide polymer, a process for their separation and their use as vehicling agents in the pharmaceutical, diagnostic, and agroalimentary industry field.

27

Rapid Expansion of Supercritical Solutions Technology

Production of Fine Particles of Steroid Drugs

Paolo Alessi, Angelo Cortesi, Ireneo Kikic, and Fabio Carli

1. Introduction

Particle size is a key factor for the performance in the use of different organic and inorganic materials. The first observation on the possibility of obtaining ultrafine powders through supercritical fluid (SF) processing was made in 1876 *(1)*, but not until 1984 did Krukonis *(2)* demonstrate the potential of SFs for processing a variety of solids that are difficult to comminute. The great advantage of using SFs is in the possibility of producing solid phases with unique morphology at mild operating conditions.

Three main processes for particle size formation with supercritical fluids are used: the SF antisolvent process or SAS (*see* Chapter 26), expansion from gas saturated solutions (PGSS), and rapid expansion of supercritical solutions (RESS). In the case of SAS *(3–8)*, the material from which particles will be formed is initially dissolved in a common solvent (an inorganic or organic compound). The supercritical solvent is added to the solution giving, as a consequence, a large variation (decrease) of the solution density. This effect leads to the reduction of the solubility of the solute that will precipitate. In the PGSS and RESS techniques, the binary system (solute and supercritical fluid) at a given temperature and pressure, is unstable, originating a two-phase system. In the PGSS process *(9)*, at temperature higher than the melting point of the solute, these two phases are liquids: the first phase is the SF saturated with the solute, the second one is constituted by the organic solute saturated with the SF. The solute-rich phase is then depressurized through a nozzle. In the RESS

From: *Methods in Biotechnology, Vol. 13: Supercritical Fluid Methods and Protocols*
Edited by: J. R. Williams and A. A. Clifford © Humana Press Inc., Totowa, NJ

process, the equilibrium between solute and supercritical fluid is reached at temperatures well below the melting point of the solute so that the two phases are the solid phase (almost pure solute) and the supercritical phase (in which the solute is partially dissolved). The dilute supercritical phase is then depressurized to allow the precipitation of solid particles.

The three processes can be considered as complementary. The choice of the process can be roughly done on the basis of solubility considerations. If the solubility of the material of interest in the supercritical fluid is higher than a few milligrams per gram of solvent, the RESS process can be used, but if the solubility is lower, the SAS process is preferred. Finally, PGSS can be used in the case of low melting point and relatively thermally stable materials.

The effectiveness of many drugs is extremely sensitive to their size and total surface area, as these strongly affect their rate of dissolution in the body. Supercritical fluids are therefore used for improving these properties in drugs. In this work, the RESS technique is described for the micronization of two steroid drugs, progesterone (*see* **Note 1**) and medroxyprogesterone acetate (*see* **Note 2**), from carbon dioxide supercritical solution. The first necessary step, to determine if the RESS technique is applicable, is the measurement of the drug solubility in the supercritical fluid and so the experimental procedure for this is included.

2. Materials

1. A carbon dioxide supply and monitoring system that supplies both the solubility measurement device and the RESS device. This is shown schematically on the left-hand side of **Fig. 1** *(10)*, and consists of a cylinder of liquid CO_2, a cooler to maintain the CO_2 as liquid in the pump (*see* **Note 3**), a pump (*see* **Note 4**), and valves, V1 and V2, of which the latter is a 3-way valve, and a filter F1 (*see* **Note 5**).
2. A system for measuring solubility, shown in **Fig. 1A**, which consists of a temperature controlled air oven (*see* **Note 6**); connecting tubing, preheating coils and a solute column (*see* **Note 7**); a regulating micrometering valve (*see* **Note 8**); a glass damper to remove fluctuations in flow rate; a mass flow meter (*see* **Note 9**); a temperature and a pressure transducer (*see* **Notes 10** and **11**); and a valve, V3, and filters F2-F4 (*see* **Note 5**).
3. A system for carrying out RESS, shown in **Fig. 1B**, which consists of a water bath with controlled heater; connecting tubing, preheating coils and a solute column (*see* **Note 7**); a bypass for the solute column with a micrometering valve, V6; heating tape with a heating controller (*see* **Note 12**); a thermostatted crystallizer with nozzle (*see* **Note 13**); a water cooler and heater for the crystallizer (*see* **Note 14**); CO_2 and N_2 cylinders connected to the crystallizer via a micrometering valve, V8; a regulating micrometering valve (*see* **Note 8**); a glass damper to remove fluctuations in flow rate; a mass flow meter (*see* **Note 9**); a temperature and two pressure transducers (*see* **Notes 10** and **11**); and valves, V4 and V5, and filters F5–F7 (*see* **Note 5**).

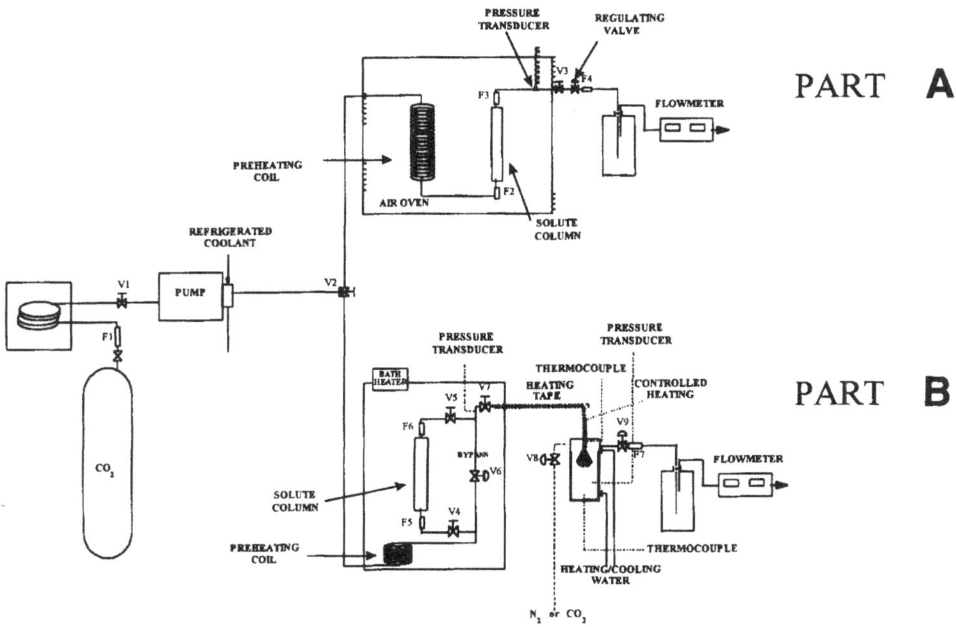

Fig. 1. Experimental apparatus.

4. A balance (a Mettler H31 balance was used).
5. A mercury porosimeter (*see* **Note 15**).
6. A differential scanning calorimeter (DSC) (*see* **Note 16**).
7. Progesterone (purity of 99.5%) (obtained from Upjohn, MI) (*see* **Note 1**).
8. Medroxyprogesterone acetate (purity of 99.5%) (obtained from Farmitalia Carlo Erba, Milan, Italy) (*see* **Note 2**).
9. Carbon dioxide—(purity of 99.98%) (obtained from SIAD, Italy).

3. Method

1. If not known, measure the solubility of the drug using the experimental system shown in **Fig. 1A**. The CO_2 is pumped as liquid into the preheating coil contained in the temperature-controlled air oven where it reaches supercritical conditions. The supercritical carbon dioxide passes into the solute column packed with the steroid and plugged at each end with glass wool to eliminate entrainment. The valve V3 is closed so the system is maintained at the experimental conditions of pressure and temperature for one night for the attainment of the thermodynamic equilibrium. The outgoing supercritical solution is then flashed to atmospheric pressure through V3 and the regulating valve, causing solute precipitation within the valve and the following filter, F4 (*see* **Note 17**). The gas is released through a glass damper, and then passes into the mass flow meter for determination of the volume and therefore the mass of CO_2. The mass of solute, trapped in the

valve and the filter, is then weighed by difference and the solubility obtained from the masses of CO_2 and solute.
2. Carry out RESS experiments using the experimental system shown in **Fig. 1B**. By means of the cooled pump, the desired pressure is reached and the carbon dioxide flows in the solute column. Part of the supercritical solvent can bypass the extraction unit by partially opening valve V6 so that different solute concentrations can be obtained. The supercritical solution is connected through the thermostatted tube to the crystallizer. A preheating controlled temperature is made also before the nozzle (*see* **Notes 18** and **19**). The cylindrical crystallizer is thermostatted to maintain constant post-expansion temperature (*see* **Note 20**). The pressure in the crystallizer is maintained at the desired value with the aid of the valve V8 through which, to avoid solvent condensation and to control the pressure, nitrogen or carbon dioxide can be injected (*see* **Note 21**). The flow rate is then measured in the flow meter after the filter F7.
3. Carry out particle characterization as follows. The morphology of the particles obtained is studied through optical microscopy and mercury porosimetry and the Mayer and Stowe model *(11)* is used for the calculation of average and particle size distribution. According to this model, cumulative percentage volume oversize distribution and differential volume distribution can be evaluated *(12)*. Specific surface area is calculated with the Rootare and Preazlow method *(13)*. DSC measurements, both on the starting material and on the processed material, are performed to assure that no structure modifications are introduced after the expansion.

4. Notes

1. Progesterone is an intermediate product for the biosynthesis of other steroid hormones, such as testosterone and estrogen. It is essential for pregnancy to occur and to evolve correctly, and it shows positive effects on protein transport and electrolytic balance. It can be taken orally or injected both for the minimization or the maximization of biological effects of endogenous hormones, for increasing insufficient production, for the correction hormone balance equilibrium, and as contraceptive.
2. Medroxyprogesterone acetate (a synthetic derivative of progesterone) has a better contraceptive action due to the fact that it is not deactivated and it has a more prolonged action *(14,15)*.
3. Such as a HAAKE K cryostat.
4. Such as an ISCO syringe pump 260D, with maximum pressure of 7500 psi.
5. Suitable valves for V1–V9 are Whitey 1/8-inch ball valves, supplied by Swagelok. These are checked to 175 bar, but were used by us up to 250 bar. Suitable filters for F1–F7 are Nupro in-line filters with 1/8-inch Swagelok tube connectors, containing AISI 316 filter elements for F4 and F7 of 0.2 μm, and 2 μm for the other filters.
6. Such as a windowed Memmert ULE 500 with a temperature range of 5°C to 220°C.
7. Stainless steel tubes, 1/8-inch outer diameter (AISI 316) are used for the connections and for the preheating coils. The preheating coils had a length of about 1 m. Solute columns were 1/2-inch outer diameter AISI 316 tubes, with a length of 25 cm.

8. For regular use, AISI 316 Whitey in-out 1/4-inch Swagelok SS-21RS4 are suitable as micrometering valves. For systems with higher solubility in carbon dioxide, SS-31RS4 micrometering valves are more suitable.
9. Such as a Bronkhorst Hi-tech EL-Flow F-111C-HA-22-V-MFM with maximum flow of 500 mL/min, calibrated for carbon dioxide.
10. Temperature transducers of precision ±0.1 K and temperature range of –70°C to 400°C, such as platinum thermocouples, Delta Ohm HD9214 and probes class 1/3 DIN TP93C.
11. Pressure transducers of precision ±0.1 bar operating up to 350 bar, such as DRUCK, PDCR 910, DPI 260.
12. Such as a 1/16 DIN Microprocessor, Watlow auto tuning control series 965.
13. A crystallizer can be made from a stainless steel cylinder (AISI 316) with an internal diameter of 4.5 cm and an internal height of 18.5 cm. This is fitted with a nozzle consisting of a stainless steel disk with a hole of diameter 30 μm or 100 μm, obtained using a laser technique.
14. Such as a HAAKE DC3 circulator and a HAAKE K cryostat.
15. Such as a Carlo Erba model 2000.
16. Such as a Perkin Elmer DSC 7.
17. To realize properly the supercritical flow is a very delicate task: the on-off valve V3 has to be opened really slowly to fill the connecting tube between V3 and the closed micrometering valve without any pressure drops in the apparatus. In fact, if the pressure drop is more than 2 bar before starting the experiment (i.e., before opening the micrometering valve), almost three hours are needed for obtaining a new equilibrium state.
18. The conditions of temperature and pressure prior to the nozzle (preexpansion conditions) are also RESS parameters: they were maintained constant in this work. It is necessary to consider the solubility behavior of the progesterone in the pressure range (90 – 240 bar) at the temperatures considered (40°C and 60°C): in these conditions progesterone, at fixed pressure, shows higher solubility values at lower temperature (it is below the crossover region). For this reason, the nozzle was maintained at lower temperature in order to avoid solute precipitation before the expansion.
19. The nozzle sizes were 100 and 30 μm. In all the experiments the lower nozzle diameter produced a lower particle size (*see* **Fig. 2**). The particle size differences correspond to large cumulative surface area variations for the samples collected. In **Fig. 3**, there is an example of the different cumulative area distribution obtained. It is important to note the possibility of obstruction of the smaller nozzle for steroids of high solubility in carbon dioxide.
20. In all the experiments an increase of particle size was observed by increasing the post-expansion temperature from 313.1 to 333.1 K.
21. The post-expansion pressure had a big influence on the size and the form of the particles obtained. Reducing the post-expansion pressure from 50 bar, caused the average particle size to decrease and the morphology to change from a dendrite structure to a prismatic one. The results for medroxyprogesterone acetate are

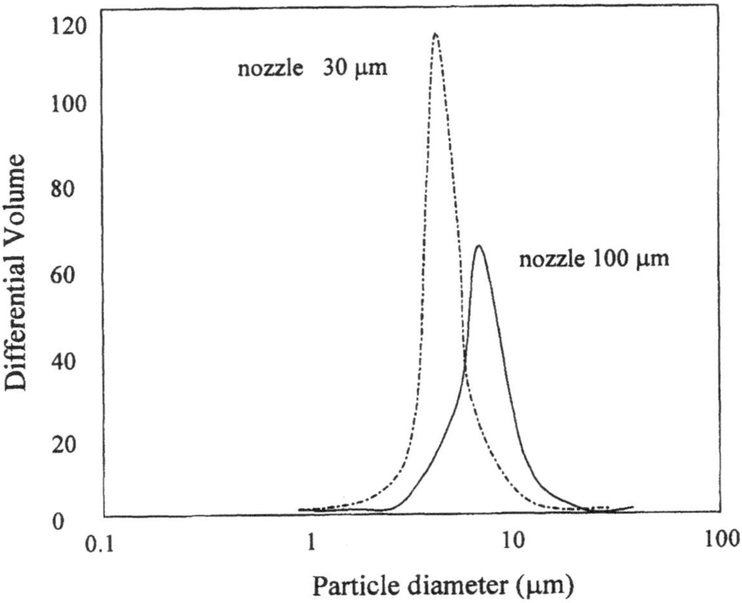

Fig. 2. Particle size distribution obtained with nozzle diameters of 30 and 100 μm. (From **ref.** ***10***, with permission).

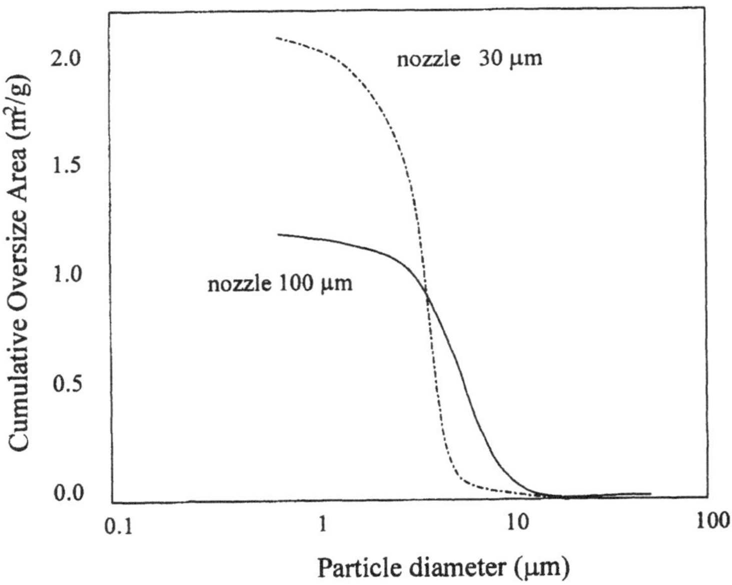

Fig. 3. Cumulative surface area distribution obtained using nozzle diameters of 30 and 100 μm. (From **ref.** ***10***, with permission).

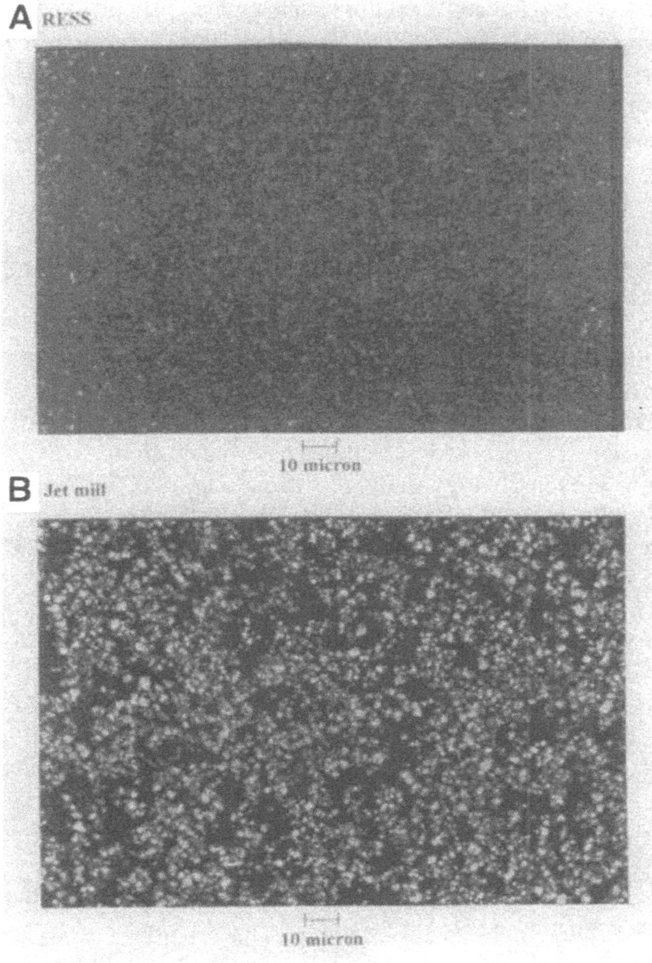

Fig. 4. Particles of medroxyprogesterone acetate micronized by **(A)** RESS (saturation pressure 150 bar, saturation temperature 60°C, pre-expansion temperature 38°C, post-expansion pressure 1 bar, post-expansion temperature 40°C, nozzle diameter 30 µm) and **(B)** by jet milling. (From **ref.** ***10***, with permission).

given in **Fig. 4** where the comparison between the RESS micronization and the traditional jet milling technique is also shown.

References

1. Hannay, J. B. and Hogarth, J. (1879) On the solubility of solids in gases. *Proc. R. Soc. Lond.* **A29,** 324.

2. Krukonis, V. (1984) Supercritical fluid nucleation of difficult to comminute solids. *Proceedings of the AIChE Meeting, November 1984, San Francisco*, Paper 140f.
3. Gallagher, P. M., Coffey, M. P., Krukonis, V. J., and Klasutis, N. (1989) Gas antisolvent recrystallization: new process to recrystallize compounds insoluble in SCF, in *Supercritical Fluid Science and Technology, ACS Symp. Series 406*, American Chemical Society, Washington, D.C., p. 334.
4. Chang, C. J. and Randolph, A. D. (1990) Solvent expansion and solute solubility predictions in gas-expanded liquids. *A.I.Ch.E. J.* **36,** 939.
5. Dixon, D. and Johnston, K. P. (1991) Molecular thermodynamics of solubilities in gas-antisolvent crystallization. *A.I.Ch.E. J.* **37,** 1441.
6. Yeo, S.-D., Lim, G.-B., Debenedetti, P. G., and Bernstein, H. (1993) Formation of microparticulate protein powders using a supercritical fluid antisolvent. *Biotechnol. Bioeng.* **41,** 341.
7. Yeo, S.-D., Debenedetti, P. G., Radosz, M., and Schmidt, H.-W. (1993) Supercritical antisolvent process for substituted para-linked aromatic polyamides: phase equilibrium and morphology study. *Macromolecules* **26,** 6207.
8. Bleich, J., Muller, B. W., and Wassmus, W (1993) Aerosol solvent extraction system: a new microparticle production technique. *Int. J. Pharm.* **97,** 111.
9. Weidner, E., Knez, Z., and Novak, Z. (1994) PGSS (particle from gas saturated solutions): a new process for powder generation. *Proceedings of the Third Int. Symp. on Supercritical Fluids, Strasbourg*, **3,** 229.
10. Alessi, P., Cortesi, A., Kikic, I., Foster, N. R., MacNaughton, S. J., and Colombo, I. (1996) Particle production of steroid drugs using supercritical fluid processing. *Ind. Eng. Chem. Res.* **35,** 4718.
11. Mayer, R. P. and Stowe, R. A. (1965) Mercury porosimetry breakthrough pressure for penetration between packed spheres. *J. Colloid. Sci.* **20,** 893.
12. Lowell, S. and Shields, J. E. (1984) *Powder Surface Area and Porosity. Powder Technology Series*, 2nd ed., Chapman and Hall, Bristol.
13. Rootare, H. M. and Preazlow, C. F. (1967) Surface areas from mercury porosimetry measurements. *J. Phys. Chem.* **21,** 2733.
14. Shacter, L., Rozenc Zeig, M., Canetta, R., Kelley, S., Nicaise, C., and Smaldone, L. (1989) Megestrol acetate: clinical experiences. *Cancer Treatment Rev.* **16,** 49.
15. Tchekmedyian, N. M., Tait, N., Moody, M., and Aisner, J. (1987) High dose megestrol acetate: a possible treatment for cachexia. *JAMA* **257,** 1195.

28

Supercritical Fluid Aerosolized Vitamin E Supplementation

Brooks M. Hybertson

1. Introduction

Rapid release of the applied pressure on a supercritical fluid solution allows the fluid to expand, its solvent strength to drop, and solute nucleation to occur, forming fine, airborne particles. This phenomenon was first observed more than 100 years ago by the scientists J. B. Hannay and J. Hogarth *(1,2)*. They released the pressure on a supercritical ethanol solution of potassium iodide and observed the precipitation of solute into a "snow" or "frost" of fine particles. This remarkable phenomenon remained largely unstudied for a century after the original findings were reported *(3)*. In 1984, Krukonis *(4)* described the precipitation of a wide variety of solute compounds by rapid expansion of supercritical fluid solutions. A subsequent review by Tom and Debenedetti *(5)* indicates that in recent years there has been increasing interest in supercritical fluid expansion processes for the formation of fine particles.

Numerous studies have examined the possibility of using supercritical fluid solution expansion processes for the manufacture of fine powders of ceramic, organic, polymeric, or pharmaceutical compounds *(3–15)*. In a typical procedure, supercritical fluid solutions are allowed to expand through an orifice, precipitated solute particles are collected by impaction on filters or surfaces, and the comminuted product is removed for use in subsequent applications. Additionally, supercritical fluid solution expansion processes have been coupled with induction of chemical reactions for the deposition of thin films *(16,17)*.

In our research, we have found that the formation of airborne solute particles by supercritical fluid solution expansion can be used *directly* as a method of drug aerosol formation and delivery. In this process, a pharmaceutical compound is

From: *Methods in Biotechnology, Vol. 13:Supercritical Fluid Methods and Protocols*
Edited by: J. R. Williams and A. A. Clifford

dissolved in a supercritical fluid, aerosolized by rapid expansion of the solution through a valve or nozzle, and directly administered to a subject via inhalation.

Drug delivery to the lungs via inhalation of pharmaceutical aerosols is extensively used for the treatment of pulmonary disorders, and may also have utility for the systemic distribution of drugs in patients *(18–23)*. Optimal pulmonary deposition of airborne drug particles in humans occurs when the particles have diameters of around 1–3 μm *(22)*. Larger particles tend to strike surfaces and deposit in the throat or upper airways, resulting in diminution of the drug's intended effect, and smaller particles may either diffuse to lung surfaces and deposit, or remain airborne and then be lost by exhalation.

Carbon dioxide is an agreeable solvent for supercritical fluid aerosolization and delivery of pharmaceuticals due to its reasonably low critical temperature (31°C) and critical pressure (1072 psi), low chemical reactivity, low cost, and well-characterized and relatively safe physiological properties.

Vitamin E, the activity of which is accounted for predominantly by the compound α-tocopherol in vivo, protects cell membranes against lipid peroxidation reactions *(24–28)*, and, as a consequence, may be protective against acute, oxidative lung injury *(29–31)*. Vitamin E is an essential nutrient, which has, of course, good bioavailability following enteral administration of food or of vitamin supplements. The process of absorption and systemic distribution via the gastrointestinal tract is very slow, however, taking on the order of days to weeks. Direct administration of vitamin E to the lungs as an inhaled aerosol is attractive because it constitutes a rapid and potentially homogeneous method for pulmonary antioxidant supplementation. Vitamin E is a good candidate for pulmonary administration using rapid expansion of a supercritical fluid solution because it is soluble in supercritical CO_2 *(32,33)*, and because rapid expansion of a supercritical CO_2 solution of vitamin E yields respirable vitamin E droplets *(34)*.

An increased oxidant burden exists in patients with acute lung injury (adult respiratory distress syndrome [ARDS]) *(35)*. Using an animal model of ARDS, we have found that lungs deficient in vitamin E are *more susceptible* to oxidative injury, and lungs supplemented with vitamin E by inhalation of the aerosol formed by rapid expansion of a supercritical CO_2 solution are *less susceptible* to oxidative injury *(36,37)*. Vitamin E deficiency potentiates oxidative damage in other models of lung injury, including exposure to ozone, hyperoxia, nitrogen dioxide, smoke, and paraquat *(25,38–43)*.

It has been shown that vitamin E can also be delivered directly to the lungs by intratracheal instillation of a liposomal formulation containing α-tocopherol *(44–47)*. However, inhalation of vitamin E droplets generated by supercritical fluid aerosolization is less invasive than endotracheal intubation and instillation of a solution containing liposomes. Furthermore, inhalation of 0.3–3 μm aero-

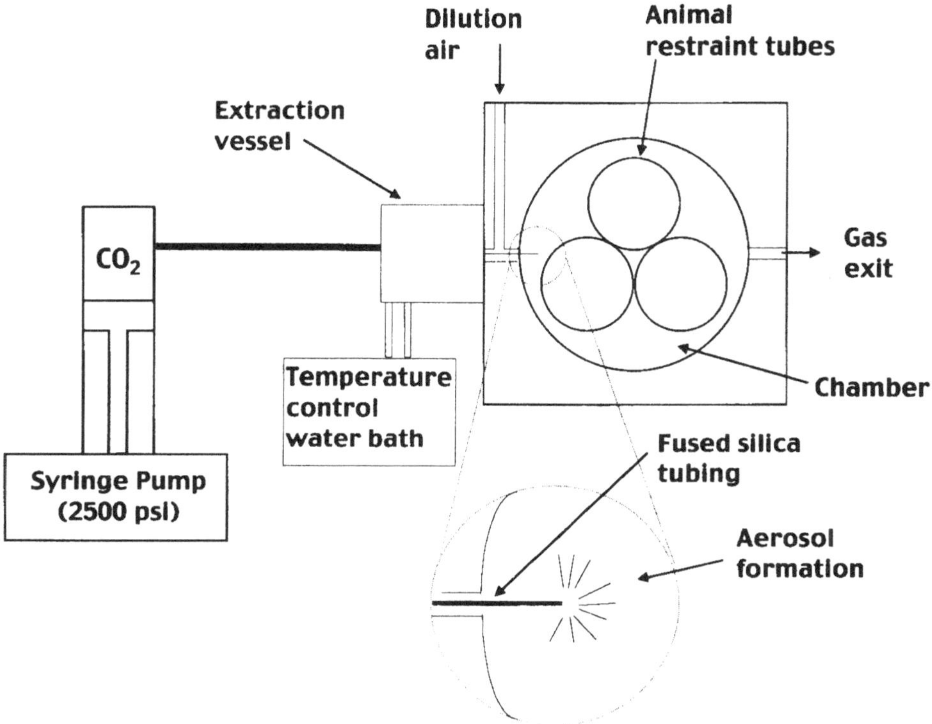

Fig. 1. Side view of nose-only exposure chamber block and rat confinement tubes.

sol particles may likely yield a more homogeneous deposition of α-tocopherol than intratracheal instillation of the larger liposomes.

A simple experimental protocol is presented in this chapter for the generation and administration of aerosolized vitamin E using supercritical fluids. In our experiments, this process yielded airborne vitamin E droplets of approx 0.7 to 2 μm in diameter, and we observed a pulmonary deposition rate in rats of approx 4 μg/min. This technique is, of course, adaptable to the extant equipment in other laboratories, and is applicable for aerosol delivery of other drug compounds that are soluble in supercritical fluids.

2. Materials

1. The aerosol generation and exposure systems for nose-only exposure. The system was constructed for delivery of vitamin E to rats (**Figs. 1** and **2**). An aluminum, high-pressure solution cell (12.7 cm × 5 cm × 3.8 cm, with internal volume approx 5 mL) was specially constructed and mounted on the outside of an acrylic exposure chamber so that the end of the nozzle (fused silica tubing, 25 μm I. D., approx 5 cm long) was inside the chamber (13.7 cm diameter × 7.3 cm deep).

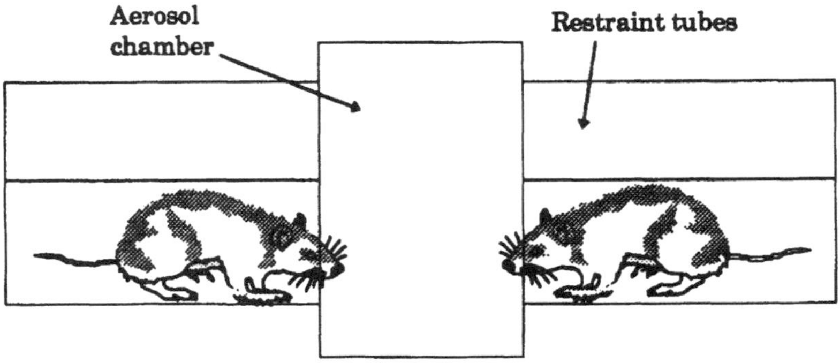

Fig. 2. End view of nose-only exposure chamber block and rat confinement tubes.

Before filling it with supercritical CO_2 for the first time, the solution cell was pressure-tested with water. A syringe pump (ISCO model 260D, Lincoln, NE) was filled with SFC-grade CO_2 and used to pressurize the cell. An air inlet was drilled into the chamber block, concentric with the restrictor nozzle. In this nose-only configuration, acrylic tubes for animal restraint (6.4 cm diameter × 25.4 cm long) were taken from a Walton horizontal smoking machine (model CTR, Process & Instruments, Brooklyn, NY), and three tubes were mounted on each side of the exposure chamber. A plunger was placed behind each rat to secure the animals during the experiment. The rats fit loosely enough inside the tubes to allow normal respiration, and their noses protruded into the exposure chamber.

2. The aerosol generation and exposure systems for whole-body exposure. The whole-body aerosol exposure chamber for rats was constructed using an acrylic chamber and an aluminum oven with a hinged top outfitted with cartridge heaters and a temperature controller (Omega Engineering, Stamford, CT) (**Fig. 3**). The oven was filled with water and used to heat a stainless steel supercritical fluid extraction vessel (Keystone Scientific, Bellafonte, PA) to 45°C. This temperature was slightly higher than that used in the nose-only system, and was chosen to ensure that the extraction vessel stayed above the critical temperature (31°C), even if some cooling occurred because of expansion of the pressurized fluid. The efflux from the extraction vessel passed through a nozzle (model no.15-12AF1 stainless steel valve, High Pressure Equipment Company, Erie, PA), and the spray from the nozzle was directed through an opening in the top of the exposure chamber.
3. A laser light-scattering particle counter such as a Particle Measuring Systems, Lasair Model 310 (Boulder, CO) *(48)*.
4. Tissue homogenizer such as Virtishear (Virtis, Gardiner, NY).
5. A centrifuge.
6. Sample vials.
7. Fluoropolymer syringe, filters 0.45-µm pore-size, such as ACRO LC13 (Gelman Sciences, Ann Arbor, MI).
8. An HPLC system consisting of an HPLC (e.g., model 510, Waters, Milford, MA),

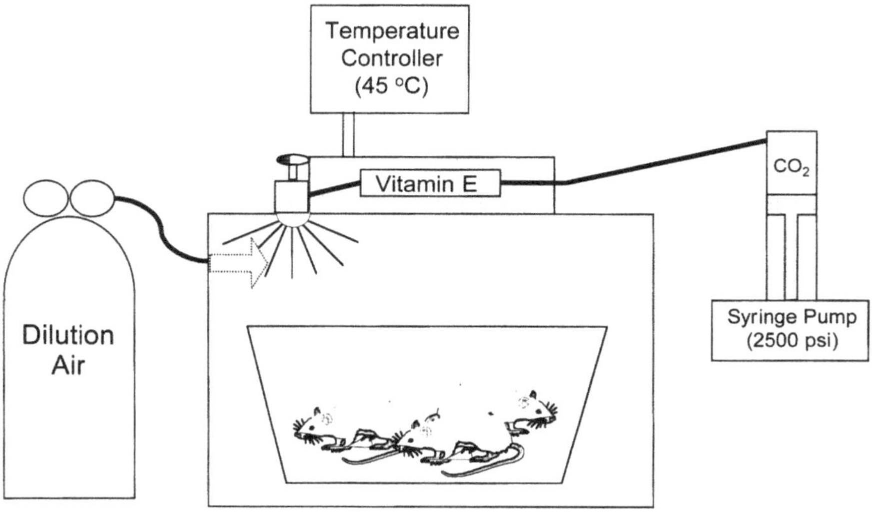

Fig. 3. Schematic of whole-body vitamin E aerosol exposure system.

a variable-wavelength UV-visible detector (e.g., Waters model 481), and HPLC software (e.g., EZChrom Chromatography Data System, San Ramon, CA).

9. Autosampler (e.g., Varian 9095).
10. HPLC C_{18} column (e.g., Nova Pak C_{18}, 15 cm long × 3.9 mm internal diameter, 5 µm particle size, Waters).
11. A short precolumn (e.g., Guard-Pak Resolve C_{18}, Waters).
12. Vitamin E (DL-α-tocopherol, 95%) was purchased from Sigma Chemical Co. (St. Louis, MO).
13. Ketamine hydrochloride and xylazine were purchased from Parke-Davis (Morris Plains, NJ) and Haver (New York, NY), respectively.
14. Hexane and HPLC-grade methanol were purchased from Fisher Scientific (Fair Lawn, NJ).
15. Supercritical fluid chromatography (SFC) grade CO_2 was obtained from Scott Specialty Gases (Plumsteadsville, PA).
16. Butylated hydroxytoluene (BHT).
17. Male Sprague-Dawley rats (250–400 g) were purchased from Sasco (Omaha, NE), were provided with water and a normal diet (Prolab Animal Diet 3000, Agway, Syracuse, NY), and were kept under appropriate conditions in accordance with the guidelines of the University of Colorado Health Sciences Center Animal Care and Use Committee.

3. Method

1. Aerosol administration of vitamin E, using the nose-only system, is carried out as follows. The extraction vessel is loaded with approximately 0.5 g vitamin E before each experiment and is packed with glass beads (3 mm diameter) to

increase the surface area and thereby increase the vitamin E dissolution rate. The vessel is maintained at 35.0 ± 0.1°C with a circulating water bath. The syringe pump is filled with CO_2 and used to pressurize the vessel. Air (6 L/min) is added through the inlet to dilute the CO_2 levels to approximately 3%. Vitamin E aerosol administration is performed with the supercritical fluid extraction vessel maintained at 2800 psig, giving a flow rate through the 25 µm internal diameter capillary tubing "nozzle" of approximately 1 mL/min of pressurized fluid. Typically, aerosol delivery is conducted on rats for 10 min with a vitamin E/CO_2 (g) concentration of approx 4–7 µg/mL (*see* **Notes 1–6**).

2. Aerosol administration of vitamin E, using the whole-body system, is carried out as follows. The stainless steel extraction vessel is loaded with 0.5 g vitamin E before each experiment and then filled with supercritical CO_2 using the syringe pump. The high pressure valve is opened until the syringe pump flow rate is 2 mL/min to maintain the pressurized fluid at 2500 psig. This pressure is slightly lower than is used in the nose-only experiments, which is chosen to give a desired flow rate. The pressure drop across the nozzle caused expansion of supercritical CO_2, loss of solvent strength, and precipitation of airborne vitamin E droplets. Air (12.5 L/min) is added to the exposure chamber to dilute the CO_2 gas in the chamber to about 3%. Rats are placed in a cage inside the exposure chamber and allowed to inhale the supercritical fluid aerosolized vitamin E droplets for 10–30 min. In sham control experiments, rats are placed inside the chamber and exposed to the same CO_2 and air without any vitamin E (*see* **Notes 1–6**).
3. Airborne droplets of supercritical fluid aerosolized vitamin E are analyzed using the laser light-scattering particle counter (*see* **Note 7**) ***(48)***. The sampling inlet is positioned inside the exposure chamber and the aerosol is sampled at a rate of 1 ft^3/min. Exposure chamber background counts are determined without aerosol generation and are subtracted to determine the vitamin E droplet size distribution.
4. Measurement of the pulmonary deposition of vitamin E aerosol is carried out as follows. Rats are subjected to aerosolized vitamin E for 30 min, anesthetized with ketamine (90 mg/kg, ip) and xylazine (7 mg/kg, ip), and then ventilated with room air via tracheostomy. After the chest is opened and the lungs are perfused blood-free with phosphate-buffered saline, the lungs are then removed, dissected free from the heart, connective tissue, and major airways, gently blotted, and then frozen. Afterward, lungs are assayed for vitamin E content by HPLC ***(26,49)***. Briefly, lung samples are weighed, and then homogenized in 1.5 mL absolute ethanol and 1.5 mL 10% ascorbic acid solution with a tissue homogenizer at maximum speed for two 30-s bursts. Samples are kept on ice during homogenizing and at all other times before HPLC analysis. Then 3 mL of hexane is added to the homogenized samples with 0.037 wt% butylated hydroxytoluene (BHT) added to prevent oxidation and increase recovery of extracted vitamin E ***(50)***. The samples are mixed by vortex, and the resultant emulsions are centrifuged (9000*g*, 5–10°C, 10 min); 2 mL of the hexane (upper) phase is withdrawn and transferred to a new tube. Hexane extracts are evaporated to dryness under flowing N_2 gas and then redissolved in 500 µL of methanol. Methanol solutions are

filtered into sample vials through 0.45 μm pore-size fluoropolymer syringe filters. Subsequently, 10 μL aliquots of the filtered samples are injected for reversed-phase HPLC analysis. A 99% methanol–1% water mobile phase is used with a flow rate of 1.0 mL/min. The detector absorption wavelength is set at 292 nm and the data collected and analyzed using HPLC software. A calibration curve is generated using standard solutions of vitamin E in methanol and used to calculate vitamin E concentrations in the lung samples.

4. Notes

1. The pressure, temperatures, flow rates, and aerosol exposure times described are representative of those used in our laboratory, but are not intended to be exclusive. Each parameter can be changed to adapt to other experimental designs.
2. A variety of safety issues should be considered when supercritical carbon dioxide is used to administer respirable aerosols of vitamin E or other drug solutes. First, the equipment used should be capable of handling high pressures. Some vendors (e.g., Keystone Scientific, Bellefonte, PA) design and sell hardware designed for supercritical fluid extraction that is suitable for use in supercritical fluid aerosolization experiments. Second, consideration should be given to the toxicity of the aerosol particles and to protecting the operator from self-exposure during the experiment. The author recommends that aerosol containment be considered even by those conducting traditional supercritical fluid extraction and solute collection experiments without intentional aerosol generation. Adventitious formation of respirable, and potentially toxic, aerosols may occur. For vitamin E aerosols—not anticipated to be very hazardous—we perform the work in a fume hood. Third, after expansion of the supercritical fluid solution, gaseous carbon dioxide levels must be diluted with air before administration.
3. In these types of experiments we recommend against the use of supercritical nitrous oxide—another lipophilic solvent with low critical temperature and pressure—because it can act as an oxidizing agent with the potential for explosive chemical reactions ***(51)***.
4. Carbon dioxide cools during expansion into the gas phase, contributing to build-up of ice and precipitated solute on the nozzle and plugging or loss of aerosol by impaction. This problem is easier to address when using a high-pressure valve such as the nozzle as opposed to a linear restrictor such as 25 μm internal diameter fused silica tubing. The valve can simply be opened farther to flush the nozzle, and the surfaces can be cleaned with a cotton swab if vitamin E build-up occurs. Others have also had success using laser-drilled orifices as supercritical fluid expansion nozzles ***(11)***. Additionally, each type of nozzle can be heated to avoid ice build-up and to maintain a desired pre-expansion temperature ***(12)***.
5. For aerosol delivery to rats and other small animals, a variety of issues must be considered. First, rats are obligate nose-breathers and inhale aerosolized drugs through their nasal passages, while humans typically inhale aerosolized drugs through their mouths. Moreover, humans have much larger airways than rats. The net result is that the optimal aerosol particle size distribution for pulmonary depo-

sition is likely different for rats and humans. Second, rats can be positioned in a variety of ways for exposure to respirable aerosols: whole body exposure (a simple procedure with minimal animal handling, the rats can be left in their cages and placed in an aerosol exposure chamber), head-only exposure, nose-only exposure (decreases the possibility of aerosol deposition on the fur and ingestion by licking), and intratracheal exposure (bypasses nasal passages and obviates the possibility of licking deposited aerosol from the fur, but requires surgical exposure of the trachea). For our vitamin E exposures, the lungs are typically studied within 5 h after the aerosol deposition. This time period is too short to allow gastrointestinal absorption and systemic distribution of vitamin E, so we used either nose-only or whole-body exposures.

6. In order to set the supercritical fluid flow rate in these experiments using the readout on the syringe pump (the piston movement that is required to maintain the set pressure), we found it necessary to allow the fluid in the pump to equilibrate to room temperature before the experiment, and to ensure that there are no leaks in the system.
7. Aerosol particle size measurement is useful in these experiments to determine whether respirable particles (of the order of 1 μm diameter) are being formed. We have used a laser light-scattering particle size analyzer, but other techniques can be considered. For example, the vitamin E aerosol could be collected on the stages of a cascade impactor, and each size fraction quantitated gravimetrically or by HPLC analysis.

References

1. Hannay, J. B. and Hogarth, J. (1880) On the solubility of solids in gases. *Chem. News* **41,** 103–106.
2. Hannay, J. B. and Hogarth, J. (1879) On the solubility of solids in gases. *Chem. News* **40,** 256.
3. Paulaitis, M. E., Krukonis, V. J., Kurnik, R. T., and Reid, R. C. (1983) Supercritical fluid extraction. *Rev. Chem. Eng.* **1,** 179–242.
4. Krukonis, V. (1984) Supercritical fluid nucleation of difficult-to-comminute solids, in *Proceedings of the A.I.Ch.E. Meeting*, San Francisco.
5. Tom, J. W. and Debenedetti, P. G. (1991) Particle formation with supercritical fluids: a review. *J. Aerosol Sci.* **22,** 555–584.
6. Brand, J. I. and Miller, D. R. (1988) Ceramic beams and thin film growth. *Thin Solid Films* **166,** 139–148.
7. Chang, C. J. and Randolph, A. D. (1989) Precipitation of microsize organic particles from supercritical fluids. *A.I.Ch.E. J.* **35,** 1876–1882.
8. Debenedetti, P. G. (1990) Homogeneous nucleation in supercritical fluids. *A.I.Ch.E. J.* **36,** 1289–1298.
9. Larson, K. A. and King, M. L. (1986) Evaluation of supercritical fluid extraction in the pharmaceutical industry. *Biotechnol. Prog.* **2,** 73–78.
10. Loth, H. and Hemgesberg, E. (1986) Properties and dissolution of drugs micronized by crystallization from supercritical gases. *Int. J. Pharm.* **32,** 265–267.

11. Matson, D. W., Fulton, J. J., Petersen, R. C., and Smith, R. D. (1987) Rapid expansion of supercritical fluid solutions: solute formation of powders, thin films, and fibers. *Ind. Eng. Chem. Res.* **26,** 2298–2306.
12. Mohamed, R. S., Halverson, D. S., Debenedetti, P. G., and Prud'homme, R. K. (1989) Solids formation after expansion of supercritical fluid mixtures, in *Supercritical Fluid Science and Technology* (Johnston, K. P. and Penniger, J. M. L., eds.) American Chemical Society, Washington, D.C., pp. 355–378.
13. Tom, J. W. and Debenedetti, P. G. (1991) Formation of bioerodible polymeric microspheres and microparticles by rapid expansion of supercritical solutions. *Biotechnol. Prog.* **7,** 403–411.
14. Kwauk, X. and Debenedetti, P. G. (1993) Mathematical modeling of aerosol formation by rapid expansion of supercritical solutions in a converging nozzle. *J. Aerosol Sci.* **24,** 445–469.
15. Debenedetti, P. G., Tom, J. W., Kwauk, X., and Yeo, S.-D. (1993) Rapid expansion of supercritical solutions (RESS): fundamentals and applications. *Fluid Phase Equilibria* **82,** 311–321.
16. Hansen, B. N., Hybertson, B. M., and Sievers, R. E. (1992) Supercritical fluid transport-chemical deposition of films. *Chem. Mater.* **4,** 749–752.
17. Hybertson, B. M., Hansen, B. N., and Sievers, R. E. (1991) Deposition of palladium films by a novel supercritical fluid transport-chemical deposition process. *Mater. Res. Bull.* **26,** 1127–1133.
18. Boyes, R. N. (1989) Prospects for drug therapy via the respiratory tract, in *Novel Drug Delivery* (Prescott, L. F., and Nimmo, W. S., eds.) Wiley, West Sussex, pp. 167–175.
19. Newman, S. P. and Clarke, S. W. (1985) Aerosols in therapy, in *Aerosols in Medicine, Principles, Diagnosis, and Therapy* (Morén, F., Newhouse, M. T., and Dolovich, M. B., eds.) Elsevier, Amsterdam, pp. 289–312.
20. Washington, N., Wilson, C. G., and Washington, C. (1989) Pulmonary drug delivery, in *Physiological Pharmaceutics, Biological Barriers to Drug Absorption* (Wilson, C. G. and Washington, N., eds.) Ellis Horwood Limited, Chichester, U.K., pp. 155–178.
21. Debs, R. J., Fuchs, H. J., Philip, R., Montgomery, A. B., Brunette, E. N., Liggitt, D., Patton, J. S., and Shellito, J. E. (1988) Lung-specific delivery of cytokines induces sustained pulmonary and systemic immunomodulation in rats. *J. Immunol.* **140,** 3482–3488.
22. Hickey, A. J. (1992) Summary of common approaches to pharmaceutical aerosol administration, in *Pharmaceutical Inhalation Aerosol Technology* (Hickey, A. J., ed.) Marcel Dekker, New York, pp. 255–288.
23. Hiller, F. C. (1992) Therapeutic aerosols: an overview from a clinical perspective, in *Pharmaceutical Inhalation Aerosol Technology* (Hickey, A. J., ed.) Marcel Dekker, New York, pp. 289–306.
24. Chow, C. K., Plopper, C. G., and Dungworth, D. L. (1979) Influence of dietary vitamin E on the lungs of ozone-exposed rats: a correlated biochemical and histological study. *Environ. Res.* **20,** 309–317.

25. Chow, C. K. (1991) Vitamin E and oxidative stress. *Free Radic. Biol. Med.* **11,** 215–232.
26. Elsayed, N. M., Mustafa, M. G., and Mead, J. F. (1990) Increased vitamin E content in the lung after ozone exposure: a possible mobilization in response to oxidative stress. *Arch. Biochem. Biophys.* **282,** 263–269.
27. McCay, P. B. (1985) Vitamin E: interactions with free radicals and ascorbate. *Annu. Rev. Nutr.* **5,** 323–340.
28. Pryor, W. A. (1991) Can vitamin E protect humans against the pathological effects of ozone in smog? *Am. J. Clin. Nutr.* **53,** 702–722.
29. Bertrand, Y., Pincemail, J., Hanique, G., Denis, B., Leenaerts, L., Vankeerberghen, L., and Deby, C. (1989) Differences in tocopherol-lipid ratios in ARDS and non-ARDS patients. *Int. Care Med.* **15,** 87–93.
30. Wolf, H. R. and Seeger, H. W. (1982) Experimental and clinical results in shock lung treatment with vitamin E. *Ann. N.Y. Acad. Sci.* **393,** 392–410.
31. Richard, C., Lemonnier, F., Thibault, M., Couturier, M., and Auzepy, P. (1990) Vitamin E deficiency and lipoperoxidation during adult respiratory distress syndrome. *Crit. Care. Med.* **18,** 4–9.
32. Ohgaki, K., Tsukahara, I., Semba, K., and Katayama, T. (1989) A fundamental study of extraction with a supercritical fluid: solubilities of α-tocopherol, palmitic acid, and tripalmitin in compressed carbon dioxide at 25°C and 40°C. *Int. Chem. Eng.* **29,** 302–308.
33. Lee, J., Chung, B. H., and Park, Y. H. (1991) Concentration of tocopherols from soybean sludge by supercritical carbon dioxide. *J. Am. Oil Chem. Soc.* **68,** 571–573.
34. Hybertson, B. M., Repine, J. E., Beehler, C. J., Rutledge, K. S., Lagalante, A. F., and Sievers, R. E. (1993) Pulmonary drug delivery of fine aerosol particles from supercritical fluids. *J. Aerosol Med.* **6,** 275–286.
35. Repine, J. E. (1992) Scientific perspectives on adult respiratory distress syndrome. *Lancet* **339,** 466–469.
36. Hybertson, B. M., Leff, J. A., Beehler, C. J., Barry, P. C., and Repine, J. E. (1995) Effect of vitamin E deficiency and supercritical fluid aerosolized vitamin E supplementation on interleukin-1-induced oxidative lung injury in rats. *Free Radic. Biol. Med.* **18,** 537–542.
37. Hybertson, B. M., Leff, J. A., and Repine, J. E. (1995) Interleukin-1, oxidant-antioxidant balance, and acute lung injury, in *The Oxygen Paradox in Biology and Medicine* (Davies, K. J. A. and Ursini, F., eds.) CLEUP University Press, Padova, Italy, pp. 763–772.
38. Elsayed, N. M., Kass, R., Mustafa, M. G., Hacker, A. D., Ospital, J. J., Chow, C. K., and Cross, C. E. (1988) Effect of dietary vitamin E level on the biochemical response of rat lung to ozone inhalation. *Drug Nutr. Interact.* **5,** 373–386.
39. Fletcher, B. L. and Tappel, A. L. (1973) Protective effects of dietary α-tocopherol in rats exposed to toxic levels of ozone and nitrogen dioxide. *Environ. Res.* **6,** 165–175.
40. Taylor, D. W. (1953) Effects of vitamin E deficiency on oxygen toxicity in the rat. *J. Physiol.* **121,** 47P-48P.
41. Wender, D. F., Thulin, G. E., Smith, G. J., and Warshaw, J. B. (1981) Vitamin E affects lung biochemical and morphologic response to hyperoxia in the newborn rabbit. *Pediatr. Res.* **15,** 262–268.

42. Chow, C. K., Chen, L. H., Thacker, R. R., and Griffith, R. B. (1984) Dietary vitamin E and pulmonary biochemical responses of rats to cigarette smoking. *Environ. Res.* **34,** 8–17.
43. Block, E. R. (1979) Potentiation of acute paraquat toxicity by vitamin E deficiency. *Lun.* **156,** 195–203.
44. Suntres, Z. E., Hepworth, S. R., and Shek, P. N. (1992) Protective effect of liposome-associated alpha-tocopherol against paraquat-induced acute lung toxicity. *Biochem. Pharmacol.* **44,** 1811–1818.
45. Suntres, Z. E. and Shek, P. N. (1993) Liposome-associated α-tocopherol: protection against bleomycin-induced lung injury. *Am. Rev. Respir. Dis.* **147,** A777.
46. Suntres, Z. E., Hepworth, S. R., and Shek, P. N. (1993) Pulmonary uptake of liposome-associated alpha-tocopherol following intratracheal instillation in rats. *J. Pharm. Pharmacol.* **45,** 514–520.
47. Suntres, Z. E. and Shek, P. N. (1995) Liposomal alpha-tocopherol alleviates the progression of paraquat-induced lung damage. *J. Drug Target.* **2,** 493–500.
48. Hickey, A. J. (1992) Methods of particle size characterization, in *Pharmaceutical Inhalation Aerosol Technology* (Hickey, A. J., ed.) Marcel Dekker, New York, pp. 219–254.
49. Vatassery, G. T. and Hagen, D. F. (1977) A liquid chromatographic method for quantitative determination of α-tocopherol in rat brain. *Anal. Biochem.* **79,** 129–134.
50. Chow, F. I. and Omaye, S. T. (1983) Use of antioxidants in the analysis of vitamins A and E in mammalian plasma by high performance liquid chromatography. *Lipids* **18,** 837–841.
51. Hansen, B. N. and Sievers, R. E. (1991) Safety letter. *Chem. Eng. News* **69,** 2.

29

Extraction of Biologically Active Substances from Wood

Jeffrey J. Morrell and Keith L. Levien

1. Introduction

Wood contains a variety of materials with potential commercial value, including resins, sugars, extractives, and other compounds that represent raw materials for syntheses. Economic recovery of many of these materials poses a challenge. Many compounds can be recovered using steam or organic solvents, but both approaches have certain drawbacks that may limit their usefulness. Steaming is energy-intensive, and the resulting condensate can be contaminated with a variety of materials. Wood can absorb large quantities of an organic solvent, increasing recovery costs. In addition, volatilization of these solvents represents an ever-increasing environmental impact.

One alternative to conventional organic solvent extraction is the use of supercritical carbon dioxide with or without small amounts of cosolvent (*see* Chapter 1). Supercritical fluids (SCFs) have advantages of rapid penetration, the ability to solubilize a variety of nonpolar compounds, low cost, and safety. These characteristics have encouraged a wealth of research into various extraction processes. SCFs have been used for extracting a variety of compounds from semiporous media ***(1–6)*** and have been used to extract a number of materials from wood, including extractives and formaldehyde ***(7–11)***. In this chapter, we will describe extraction methods for recovering various materials from western juniper, Alaska cedar, and pentachlorophenol-treated Douglas fir wood chips.

Western juniper (*Juniperus communis*) has invaded significant amounts of rangeland in western North America where it consumes a disproportionate amount of water and reduces rangeland productivity. This species currently

From: *Methods in Biotechnology, Vol. 13: Supercritical Fluid Methods and Protocols*
Edited by: J. R. Williams and A. A. Clifford © Humana Press Inc., Totowa, NJ

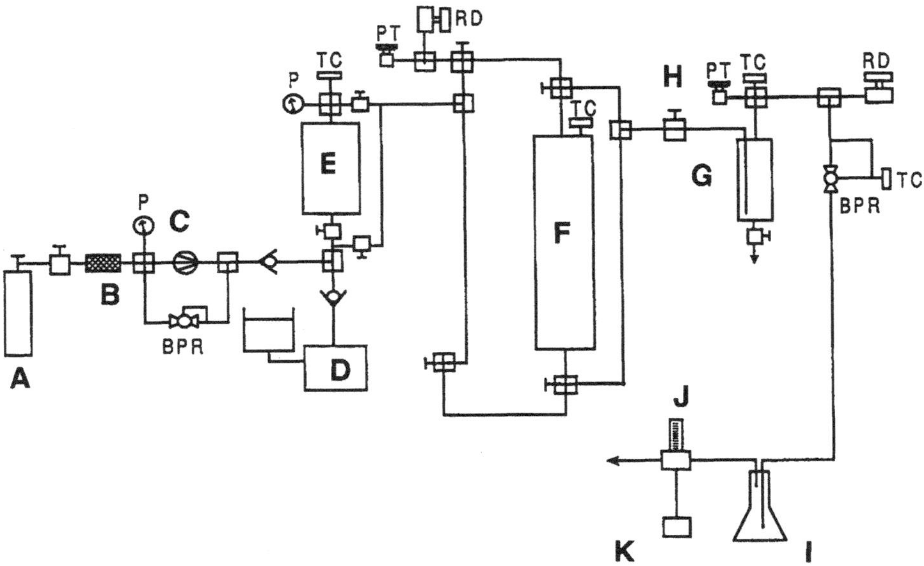

Fig. 1. Schematic of system employed for SCF extraction of western juniper, Alaska cedar, and pentachlorophenol-treated Douglas fir. A, CO_2; B, filter; C, compressor; D, cosolvent pump; E, vessel 1; F, vessel 2; G vessel 3; H, metering valve; I, cold trap; J, flowmeter; K, totalizer; BPR, back pressure regulator; P, pressure gauge; PT, pressure transducer; RD, rupture disk; TC, thermocouple.

has little or no commercial value, but other juniper species are extracted for their oils ***(12–14)***. Similarly, Alaska cedar (*Chaemacyparis nootkatensis*) contains a wealth of potential medicinal products that may be more efficiently extracted using SCF processes. At the other end of the spectrum, SCFs may also be used for extracting synthetic preservatives from wood at the end of its useful service life, making the wood safer to dispose. The prospect for using SC carbon dioxide for recovering juniper oil from western juniper, potential medicinals from Alaska cedar, and the wood preservative pentachlorophenol from Douglas fir is addressed below.

2. Materials

1. Chemicals: liquid carbon dioxide (99.9 wt%) is used in all the studies. Reagent-grade acetone and methanol are used as cosolvents. Methanol is also used for extraction of wood for analysis.
2. Treatment vessels: extractions are performed in an ISCO Series 2000 Extractor (**Fig. 1**) configured as a flow through system with a 15-mm-diameter by 54-mm-long extraction chamber.
3. Safety equipment: safety gloves and glasses.

4. Analytical equipment: a gas chromatograph equipped with a flame-ionization detector (Shimadzu GC17A, Shimadzu, Kyoto, Japan). A DBTM-S [(5% phenyl)methylpolysiloxane] column, 25 m × 0.25 µm ID (0.25 µm film thickness of liquid phase) (J and W Scientific, Folsom, CA). An ASOMA 8620 x-ray fluorescence analyzer (ASOMA Instruments, Austin, TX) with elements and filters specific for analyzing pentachlorophenol.
5. Wood: Western juniper (*Juniperus communis*), Alaska cedar (*Chamaecyparis nootkatensis*) or pentachlorophenol-treated Douglas fir sapwood (*Pseudotsuga menziesii*) ground to pass a 20 mesh screen (*see* **Note 1**).

3. Method

1. Extraction schemes can be directed at materials naturally deposited during growth of the tree or synthetic chemicals impregnated at an earlier time to protect the wood from biodegradation (*see* Chapter 30). In either instance, the goal is to select solvents, cosolvents, and extraction conditions that maximize recovery of the desired products while minimizing recovery of interfering materials.
2. A measured amount of Douglas fir, Alaska cedar, or western juniper is placed in a 15-mm-diameter by 54-mm-long stainless steel vessel, which is then placed into the extraction system.
3. Supercritical carbon dioxide, with or without a cosolvent, is then sent through the vessel at a rate of 12 mL/min for periods ranging from 15 to 60 min at 45 or 75°C and 1800, 3600, or 4500 psi (*see* **Notes 2–4**).
4. The product mixture is then depressurized and residual compounds trapped by bubbling the mixture through methanol. The weight of wood before and after treatment provides a measure of extraction efficiency.
5. The compounds present in the juniper and cedar extracts are qualitatively examined by gas chromatography ***(15)***. Compounds from juniper and Alaska cedar are quantified by diluting a 0.1 mL aliquot of extract in 4.9 mL of hexane. A 1 µL aliquot of this sample is injected into the GC followed by a 2 µL air injection, followed by 1 µL of sample. GC operating conditions are split injection system (1:50 rate), carrier gas He, flow rate 30 mL/min, hydrogen flow, 50 mL/min. Temperature programming is held at 100°C for 1 min, then increased to 150°C at 5°C/min, to 220°C at 3°C/min, and finally to 240°C at 5°C/min. The GC is held at 240°C for 2 min. Injector and detector temperatures are 250°C and total analysis time should be approx 40 min.

4. Notes

1. Particle size can have a major effect on extraction efficiency. A chip thickness of 0.5 mm or less will produce more efficient extraction (**Fig. 2**).
2. Pressure will have less effect on extraction efficiency than will extraction time for various materials (**Figs. 3–5**).
3. Cosolvents can markedly improve extraction efficiency. In general, extraction efficiency in SC-CO_2 declines with increased polarity of the material being extracted. Cosolvents such as methanol or acetone can help to overcome these difficulties.

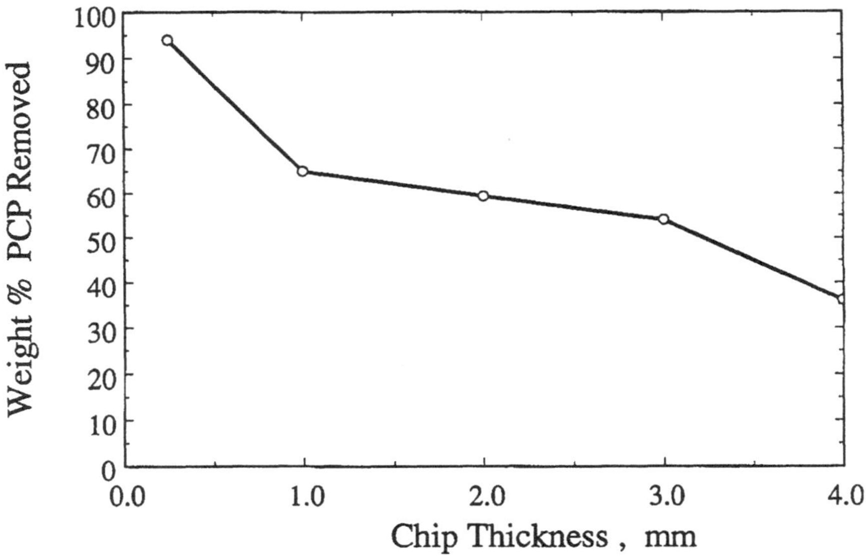

Fig. 2. Effect of particle size (chip thickness) on supercritical fluid extraction of pentachlorophenol from Douglas fir chips.

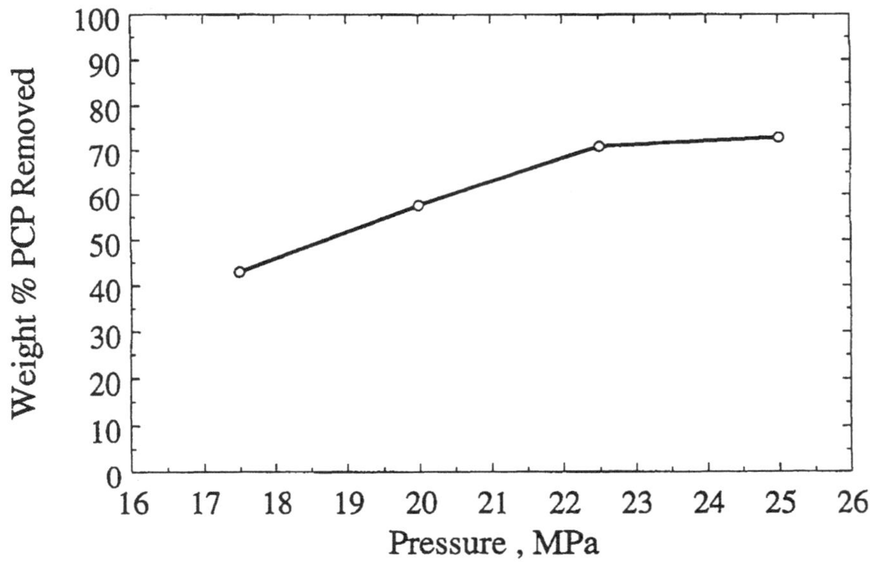

Fig. 3. Effect of pressure level on supercritical fluid extraction of pentachlorophenol from Douglas fir chips using SC-carbon dioxide at 80°C for 60 min.

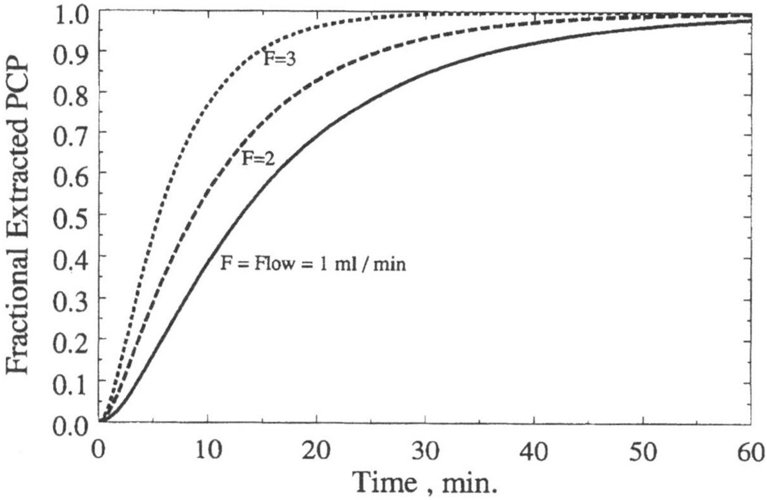

Fig. 4. Effect of extraction time on supercritical carbon dioxide extraction of pentachlorophenol from Douglas fir chips.

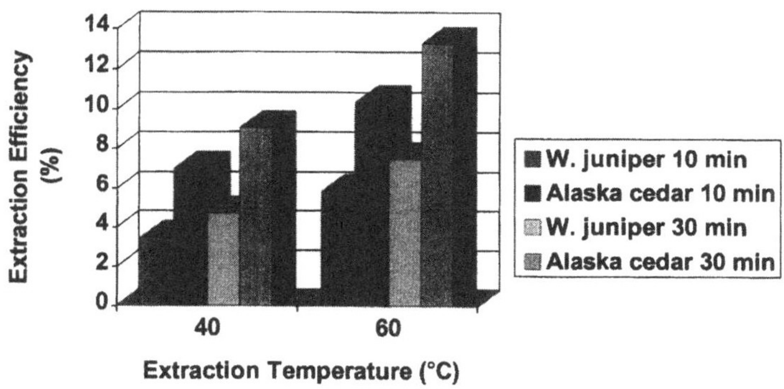

Fig. 5. Effect of extraction time and temperature on recovery of extractives from western juniper or Alaska cedar chips during supercritical carbon dioxide extraction at 13 MPa using a 10 mL/min CO_2 flow rate *(15)*.

4. Caution should be exercised when using cosolvents for extraction, since a portion of the cosolvent remains in the material following depressurization. This residual cosolvent would artificially depress extraction efficiency.

References

1. Hubert, P. and Vitzthum, O. G. (1978) Fluid extraction of hops, spices, and tobacco with supercritical gases. *Angew. Chem. Int. Ed. Eng.* **17,** 710–715.

2. Krukonis, V. J. (1988) Processing with supercritical fluids: overview and applications. *ACS Symp. Ser.* **366,** 26–43.
3. Hoyer, G. G. (1985) Extraction with supercritical fluids: why, how, and so what? *Chem. Tech.* July, pp. 440–448.
4. Larson, K. A. and King, M. K. (1986) Evaluation of supercritical fluid extraction in the pharmaceutical industry. *Biotech. Prog.* **2,** 73–82.
5. Moyler, D. A. (1993) Extraction of essential oils with carbon dioxide. *Flavour Frag. J.* **8,** 235–248.
6. Williams, D. F. (1981) Extraction with supercritical gases. *Chem. Eng. Sci.* **36,** 1769–1788.
7. Larsen, A., Jentoft, N. A., and Greibrokk, T. (1992) Extraction of formaldehyde from particle board with supercritical carbon dioxide. *For. Prod. J.* **42,** 45–48.
8. Ohira, T., Ytagai, M., Itoya, Y., and Nakamura, S. (1996) Efficient extraction of hinokitiol from wood of Hiba with supercritical carbon dioxide. *Mokuzai gakkaishi* **42,** 1006–1012.
9. Ohira, T., Terauchi, F., and Yatagai, M. (1994) Tropolones extracted from the wood of western red cedar by supercritical carbon dioxide. *Holzforschung* **48,** 308–312.
10. Ritter, D. C. and Campbell, A. G. (1991) Supercritical carbon dioxide extraction of southern pine and ponderosa pine. *Wood Fiber Sci.* **23,** 98–113.
11. Sahle Demessie, E., Yi, J. S., Levien, K. L., and Morrell, J. J. (1997) Supercritical fluid extraction of pentachlorophenol from pressure-treated wood. *Sep. Sci. Tech.* **32,** 1067–1085.
12. Adams, R. (1987) Investigation of *Juniperus* species of the United States for new sources of cedarwood oil. *Econ. Bot.* **41,** 48–54.
13. Adams, R. P. (1987) Yields and seasonal variation of phytochemicals from *Juniperus* species of the United States. *Biomass* **12,** 129–139.
14. Clark, A., McChesney, J., and Adams, R. (1990) Antimicrobial properties of heartwood, bark/sapwood, and leaves of *Juniperus* species. *Phytother. Res.* **4,** 15–19.
15. Acda, M. N., Morrell, J. J., Silva, A., Levien, K. L., and Karchesy, J. (1998) Using supercritical carbon dioxide for extraction of western juniper and Alaska-cedar. *Holzforschung* **52,** 472–474.

30

The Deposition of a Biocide in Wood-Based Material

Jeffrey J. Morrell and Keith L. Levien

1. Introduction

Wood is a unique combination of the biological polymers cellulose, lignin, and hemicellulose. In living trees, wood serves a structural function supporting the foliage in its never-ending struggle to collect sunlight for photosynthesis. The properties of wood that make it so useful to the living tree also have many applications for man.

In its native state, wood has high-strength per unit weight and is an important building material. Cellulose is a high strength polymer and dissolution of the lignin matrix surrounding other material forms the basis for the pulp and paper industry *(1)*. In addition, other materials in wood, termed extractives, play no structural role, but protect some woods from biodeterioration. A number of these compounds may have medicinal applications. Recovery of these compounds can be accomplished by steam extraction or use of nonpolar solvents. These processes can be costly, given the relatively low levels of compound present (<2% wt/wt) in many woods.

While wood has many positive attributes, it is a natural organic material and, as such, is susceptible to biological degradation under certain temperature, aeration and moisture levels *(2)*. The damage can be prevented by excluding moisture, but this is not always possible. Instead, the wood can be treated with toxic chemicals that inhibit biological attack. These processes usually involve application of a chemical preservative using a combination of vacuum and/or pressure to "force" this chemical deep into the wood. Pressure treatment of wood generally produces an envelope of preservative whose thickness varies with wood species. Some species, such as southern pine, have thick bands of permeable sapwood that readily accept treatment. Other species, such as spruces and Douglas-fir, have thin bands of sapwood, and a high percentage

From: *Methods in Biotechnology, Vol. 13: Supercritical Fluid Methods and Protocols*
Edited by: J. R. Williams and A. A. Clifford

of largely impermeable heartwood. These species are more difficult to treat. The primary factor limiting treatment is the relative permeability of the passages between individual cells. These passages, termed *pits,* vary in their permeability depending upon wood species and age (i.e., whether they are sapwood or heartwood). The effective treatment of many wood species has been the subject of considerable research, but conventional treatment processes are limited by the inability to force liquid preservative solutions across the pit membranes in the heartwood.

There is an increasing need for improved methods to effectively deliver fluids into a wood matrix for reacting, extracting, or depositing other materials. Each of these applications depends upon the ability of a fluid to move through the semiporous wood. What then can be done to overcome the inherent resistance of many wood species to fluid flow?

One approach to improve the process is to alter the characteristics of the treatment medium to increase diffusivity or reduce viscosity *(3–7)*. While conventional solvents can be altered to improve diffusivity or viscosity, the size of such changes is limited. Alternatively, substitute carriers must be identified. As an extreme, gases with high diffusivities could be used, but these fluids lack the solvating capabilities of liquids. An intermediate approach to impregnate wood-based materials is to use supercritical fluids (SCFs) as carriers. SCFs have diffusivities that are intermediate between liquids and gases, and a number of these fluids have the ability to solubilize materials at levels that can approach those of liquid carriers (*see* Chapter 1). SCFs have been explored for extraction and deposition of a variety of materials in wood *(8–17)*. In this chapter, we will describe a method for the deposition of biocides into wood.

2. Materials

1. Chemicals: liquid carbon dioxide (99.9 wt%) is used in all the studies. Reagent grade methanol is used as a cosolvent and for the extraction of wood for analysis. Tebuconazole or α-[2-(4-chlorophenyl)ethyl]-α-(1,1-dimethyethyl)-1*H*-1,2,4-triazole-1-ethanol (Bayer Inc., Pittsburgh, PA) and IPBC or 3-iodo-2-propynyl-butylcarbamate (Troy Corp., Newark, NJ) are used for biocide deposition.
2. Treatment vessels: Deposition studies are performed using a Newport Scientific Super Pressure System (configured so that SC carbon dioxide circulates over a bed of biocide in one vessel, then over wood in a 1.8-L second vessel). The SCF can then be decompressed in a third vessel and released through a trap and vent system (**Fig. 1**).
3. Safety equipment: safety gloves and glasses.
4. Analytical equipment: high performance liquid chromatograph equipped with a UV-visible detector (230 nm) and capable of gradient elution. A Shimadzu HPLC equipped with a 10 cm stainless steel column (4.6 mm ID) filled with Hypersil ODS (3 μm).
5. Wood: Douglas-fir heartwood (*Pseudotsuga menziesii*).

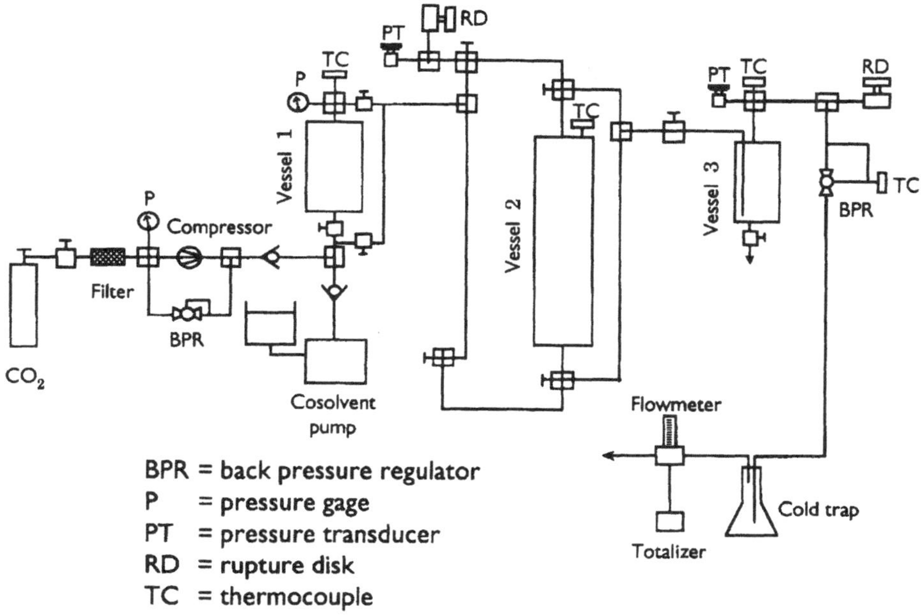

Fig. 1. Schematic of system employed for SCF impregnation of wood-based materials.

3. Method

1. There are an infinite array of possible conditions for supercritical fluid deposition of biocides into wood-based materials.
2. In our trials, we use pressures ranging from 12 to 30 MPa (1740 to 4350 psi) and temperatures from 40°C to 70°C (*see* **Note 1**). Prior studies have shown that the solubility of some of our biocides in SC carbon dioxide under these conditions is adequate for delivering sufficient quantities of chemical to confer wood protection *(18,19)*.
3. Deposition can be altered by choice and amount of cosolvent or by varying pressure or temperature, although these operating conditions have tended to be inconsistently related to retention (*see* **Notes 2** and **3**).
4. A variety of biocides can be solubilized using supercritical carbon dioxide with or without methanol as a cosolvent *(18,19)*. For example, 3-iodo-2-propynylbutylcarbamate or IPBC (Troy Chemical Co., Newark, NJ) biocide has excellent activity against many decay fungi and is widely used in North America and Europe for protecting windows and door frames from decay. IPBC has exceptional solubility in SC carbon dioxide (**Fig. 2**). Thiocyanomethylthiobenzothiazole or TCMTB (Buckman Laboratories, Memphis, TN) has also proven useful when acetone is added as a cosolvent.
5. End-sealed, small stakes of Douglas fir heartwood (38 mm × 38 mm × 100 mm long) are conditioned to a constant weight at 70 RH and 20°C.

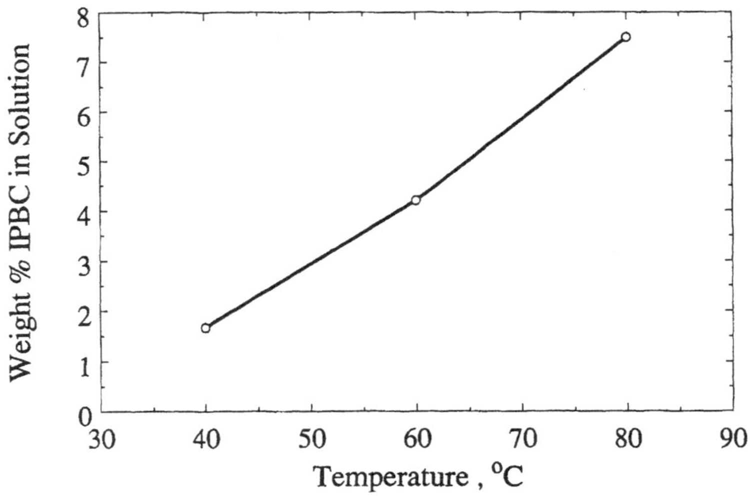

Fig. 2. Effect of temperature on solubility of IPBC in supercritical carbon dioxide at 25 *M*Pa.

6. The specimens are placed in the treatment vessel and SC carbon dioxide is circulated over IPBC or TCMTB packed on filter paper to produce a solution that is saturated with biocide.
7. The solution is then introduced into the treatment vessel where it is allowed to flow over the wood specimens. The direction of flow through the vessel is reversed at 3-min intervals (*see* **Note 4**).
8. At the end of the desired time, the pressure is rapidly released (~1000 psi/min). The resulting drop in pressure results in biocide deposition within the wood (*see* **Note 5**).
9. For analysis, 5-mm-thick sections are cut from the top, bottom, and middle of each sample. These samples are further divided into inner and outer zones. The wood from a given zone is ground to pass a 30 mesh screen before retention analysis by neutron activation analysis (NAA) for residual iodine by the Ecole Polytechnique (Montreal, Canada). In NAA, a weighed sample is irradiated, and the resulting radiation levels are measured. The induced activity is proportional to the concentration of the element of interest in the sample. Iodine is used as an indicator of IPBC retention, which is expressed by kilograms per cubic meter of wood.

4. Notes

1. When impregnating wood-based materials, certain wood species are sensitive to pressure and may collapse. These include the spruces. Slow depressurization at the conclusion of the cycle can reduce this damage.
2. Cosolvents can improve solubility of many biocides, but have relatively less effect at higher temperature with the materials evaluated.

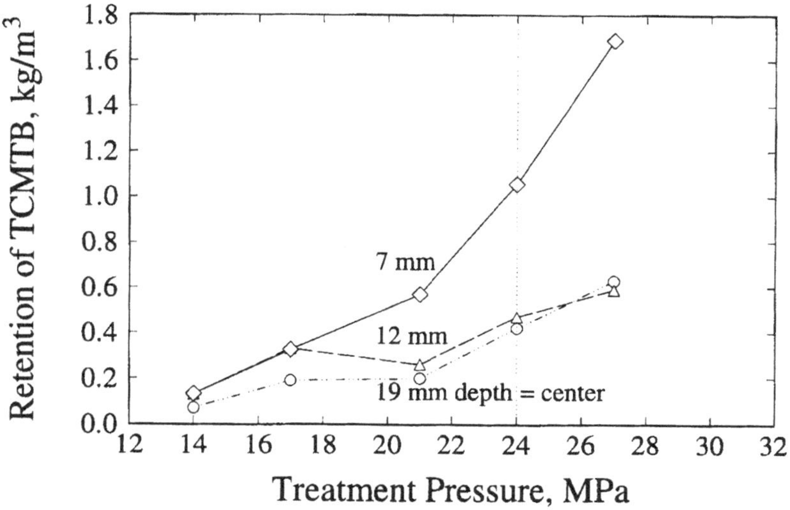

Fig. 3. Effect of treatment pressure on retention and distribution of thiocyanomethylthiobenzothiazole (TCMTB) in Douglas-fir blocks following SCF impregnation at 50°C for 30 min.

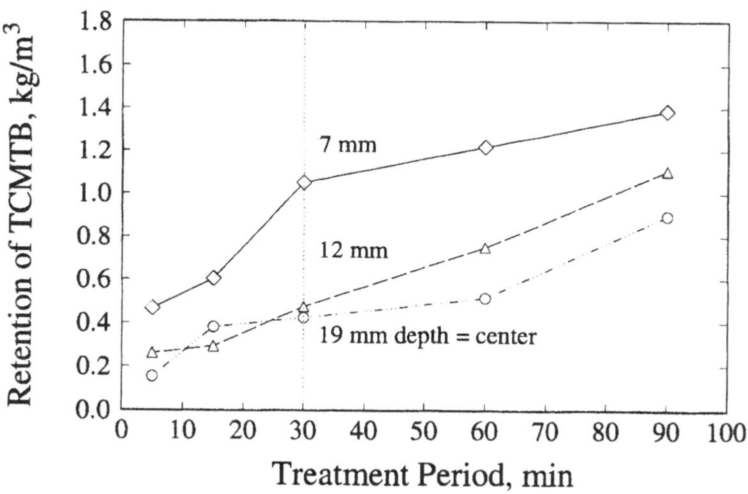

Fig. 4. Effect of treatment time on retention and distribution of TCMTB in Douglas-fir heartwood blocks following impregnation at 50°C and 24 MPa.

3. Solubility of some materials, including IPBC can be extremely high as temperature increases (**Fig. 2**). This can result in excessive consumption of pure compounds. Begin with lower pressures and temperatures to ensure that extremely

soluble biocides are not rapidly lost from the vessel or add these materials to the cosolvent so they can be metered into the fluid in a controlled manner.

4. Circulation direction and venting of the SCF can affect biocide distribution on one end of a wood sample. For best results, reverse flow through the vessel at 3- to 5-min intervals.
5. Biocide deposition is sensitive to a number of variables including treatment pressure (**Fig. 3**) or time (**Fig. 4**). The effects of these variables on impregnation remain poorly understood.

References

1. Biermann, C. (1993) *Essentials of Pulping and Papermaking*. Academic Press, New York.
2. Zabel, R. A. and Morrell, J. J. (1992) *Wood Microbiology: Decay and its Prevention*. Academic Press, New York.
3. Hoyer, G. G. (1985) Extraction with supercritical fluids: why, how, and so what? *Chem. Tech.* July, pp. 440–448.
4. Ito, N. T., Someya, T., Taniguchi, M., and Inamura, H. (1984) *Japanese Pat.* 59–1013111.
5. Kayihan, F. (1992) Method of perfusing a porous workpiece with a chemical composition using cosolvents. *U.S. Pat.* 5094892.
6. Krukonis, V. J. (1988) Processing with supercritical fluids: overview and applications. *A.C.S. Symp. Ser.* **366,** 26–43.
7. Paulaitis, M. E., Penninger, J. M. L., Gray Jr., R. D., and Davidson, P. (1983) *Chemical Engineering at Supercritical Fluid Conditions*. Ann Arbor Science, Ann Arbor, MI.
8. Acda, M. N., Morrell, J. J., and Levien, K. L. (1995) Impregnation of wood-based composites using supercritical fluids: a preliminary report. *Proc. Can. Wood Preserv. Assoc.* **16,** 9–28.
9. Acda, M. N., Morrell, J. J., Silva, A., Levien, K. L., and Karchesy, J. (1998) Using supercritical carbon dioxide for extraction of western juniper and Alaska-cedar. *Holzforschung* **52,** 472–474.
10. Larsen, A., Jentoft, N. A., and Greibrokk, T. (1992) Extraction of formaldehyde from particle board with supercritical carbon dioxide. *For. Prod. J.* **42,** 45–48.
11. Ohira, T., Terauchi, F., and Yatagai, M. (1994) Tropolones extracted from the wood of western red cedar by supercritical carbon dioxide. *Holzforschung* **48,** 308–312.
12. Ohira, T., Yatagai, M., Itoya, Y., and Nakamura, S. (1996). Efficient extraction of hinokitiol from wood of Hiba with supercritical carbon dioxide. *Mokuzai gakkaishi* **42,** 1006–1012.
13. Morrell, J. J., Levien, K. L., Sahle Demessie, E., Kumar, S., Smith, S., and Barnes, H. M. (1993) Treatment of wood using supercritical fluid processes. *Proc. Can. Wood Preserv. Assoc.* **14,** 6–25.
14. Smith, S. M., Sahle Demessie, E., Morrell, J. J., Levien, K. L., and Ng, H. (1993) Supercritical fluid (SCF) treatment: its effect on bending strength and stiffness of ponderosa pine sapwood. *Wood Fiber Sci.* **25,** 119–123.

15. Smith, S. M., Morrell, J. J., Sahle Demessie, E., and Levien, K. L. (1993) Supercritical fluid treatments: effects on bending strength of white spruce heartwood. *Int. Res. Group Wood Pres. Doc. No. IRG/WP*, 93–20008, Stockholm, Sweden.
16. Sunol, A. K. (1991) Supercritical fluid-aided treatment of porous materials. *U.S. Pat.* 4992308.
17. Sahle Demessie, E., Yi, J. S., Levien, K. L., and Morrell, J. J. (1997) Supercritical fluid extraction of pentachlorophenol from pressure-treated wood. *Sep. Sci. Tech.* **32,** 1067–1085.
18. Sahle Demessie, E. (1994) Deposition of chemicals in semi-porous solids using supercritical fluid carriers. Ph.D. Thesis, Oregon State University, Corvallis, OR.
19. Junsophonsri, S. (1994) Solubility of biocides in pure and modified supercritical carbon dioxide. M.Sc. Thesis, Oregon State University, Corvallis, OR.

31

Critical Point Drying of Biological Specimens for Scanning Electron Microscopy

Douglas Bray

1. Introduction

Although several methods can be used to dry specimens for examination with the scanning electron microscope (SEM), critical point drying (CPD) is by far the most widely used. The technique was first introduced by Anderson *(1)* to preserve three-dimensional structure of biological specimens for transmission electron microscopy. Later, it was reintroduced *(2)* as a method of obtaining dry specimens for SEM examination.

Because specimens placed in the SEM are examined under vacuum, they must first be thoroughly dried. Direct air-drying can result in considerable distortion of specimen shape due to the adverse effects of surface tension forces. Water has an extremely high surface tension (72.75 N m^{-2} at 20°C) and the receding liquid boundary results in unacceptably high levels of drying artifact even for the toughest specimens. Replacing specimen water with solvents of lower surface tension before air-drying has produced good results for some *(3,4)*, but not all specimens. An alternative method, freeze-drying *(5)*, which removes specimen water by the process of sublimation, has proven successful for some specimens, particularly small ones that can be frozen rapidly enough to prevent ice crystal formation. CPD, however, because of its applicability to virtually all specimens, remains the benchmark method against which all other procedures are compared.

CPD is based on the principle that a liquid held in a sealed chamber will simultaneously expand and evaporate when subjected to an increase in temperature. As the kinetic energy of molecules in the liquid phase increases, more of them enter the gas phase, resulting in a progressive decrease in density of the liquid and consequent increase in density of the gas. At a certain combination

From: *Methods in Biotechnology, Vol. 13: Supercritical Fluid Methods and Protocols*
Edited by: J. R. Williams and A. A. Clifford © Humana Press Inc., Totowa, NJ

of temperature and pressure, the *critical point*, the densities of both phases are equal and the boundary between them disappears, thus reducing surface tension to zero. The specific coordinates of the critical point are different for each solvent. The critical point of water is so high (374°C and 217.7 atm) that biological specimens would be cooked and destroyed under these conditions. It is necessary therefore to replace water with a *transitional solvent* that has a critical point compatible with biological specimens. Maintaining a specimen in the transitional solvent at, or above, its critical point, while gradually venting off the solvent, results in a dried specimen that has not been subjected to the deleterious effects of surface tension forces.

Although other fluids have been used in the past, the most commonly used transitional solvent is CO_2. It is readily available, inexpensive, environmentally friendly, and has a reasonable critical temperature and pressure (31.1°C and 72.9 atm). Since water and CO_2 are not miscible, a dehydrating agent that is miscible with both must be employed. Ethanol and acetone are the two most commonly used dehydrating agents. Following fixation, specimens are dehydrated using a graded series of either solvent.

Before dehydration, specimens are generally fixed to stabilize the tissue components and to reduce their extraction in the dehydration and transitional solvents. Common practice is to fix samples initially with glutaraldehyde or a combination of glutaraldehyde and formaldehyde, followed by a postfixation in osmium tetroxide and then on to the dehydration series. The postfixation step is often bypassed if the specimen has been fixed for long enough in the aldehydes. It is also possible to process specimens that have been previously preserved for light microscopy, since most of the common fixatives used, i.e., AFA (acetic acid, formaldehyde, and alcohol), Bouin's, and so on, preserve structure well enough for SEM examination. Some specimens can be adequately preserved by simply immersing them directly into boiling 95% ethanol (Dr. D. Nelson, personal communication).

There are several brands of critical point dryers on the market, all of which use the same basic methodology. For the sake of brevity, and because it is the most widely used, only the Polaron apparatus will be described here. A CPD apparatus is essentially a bomb, as it is designed to withstand high pressures. The heart of the Polaron critical point dryer (**Fig. 1**) is a thick-walled cylindrical pressure chamber surrounded by a water jacket and fitted with inlet and outlet ports. At the front end of the chamber is a 5-cm-thick glass window covered with a perspex shield through which the process can be viewed. At the back end is a thick metal door that can be unscrewed to gain entry to the pressure chamber. Pressure and temperature can be monitored by using gauges located on the top, and a safety valve that will rupture if the pressure exceeds 12.8 Pa (1850 psi) is contained in the base. Liquid CO_2 from an attached cylin-

Fig. 1. Polaron E3000 CPD apparatus: *A, B,* and *C,* the inlet, vent, and drain valve handles, respectively; *D,* door; and *W,* inspection window. Pressure *(left side)* and temperature *(right side)* gauges are indicated. Note that the apparatus is mounted on a wooden base for additional stability.

der is admitted through the inlet valve. CO_2 can be vented as a gas through the vent valve, or as a liquid, along with residual dehydrating fluid, through the drain valve at the bottom.

After dehydration to 100% is complete, specimens are placed in the boat-shaped sample carrier and transferred, along with a small amount of dehydrant, to the precooled pressure chamber of the critical point dryer. Once the door is closed, a plunger in the base of the trough opens a valve that drains excess dehydrating solvent into the bottom of the pressure chamber where the drain is located. The inlet valve is then opened to admit liquid CO_2, and when the chamber is almost full, both the vent and drain valves are opened for a short period to remove the dehydrating solvent in the base. This flushing process is repeated a number of times over a period of 0.5 to 2-h, depending on specimen size, to eliminate all traces of the dehydrating solvent. Once CO_2 infiltration is judged

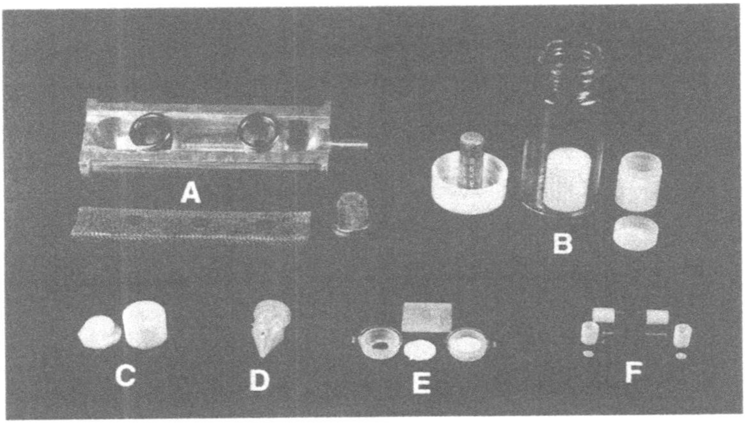

Fig. 2. Assorted specimen holders and vials used for critical point drying, processing, and storing of dried specimens. **(A)** The CPD transfer carrier, together with metal mesh vials and cover. **(B)** Scintillation vial with a desiccant capsule glued to the lid. Inside the vial is a porous plastic container (also shown uncapped beside the vial) for storing dry specimens. **(C)** A microporous specimen capsule with lid. **(D)** A conical tip BEEM capsule that has had holes drilled through the sides for fluid transfer. **(E)** A standard BEEM capsule that has been cut to form a hollow tube. Holes that can be covered with filter paper have been drilled in the lids. **(F)** Two glass tubes (one open and one closed) with plastic ends that will accept TEM grids. All containers can be purchased through most EM supply companies and then modified accordingly.

to be complete, all valves are closed, and the pressure is gradually increased by passing hot water through the water jacket. As the critical temperature and pressure are approached, the boundary line between the liquid and gas phases becomes less distinct, and finally disappears when the pressure exceeds 7.4 MPa (1073 psi). The temperature can also be monitored, but it is a poor indicator because of a lag between the water jacket temperature and that of the pressure chamber. After surpassing the critical point, the temperature is maintained well above 31.1°C, while the pressure is gradually lowered to 1 atm by slowly venting off CO_2 gas. The dried specimen, which has not been subjected to deleterious surface tension effects, can then be removed and either mounted directly, or stored in a desiccated environment for later mounting.

2. Materials

1. Specimen containers (*see* **Fig. 2**) *(6–9)*.
2. Primary fixatives such as buffered (0.1 *M*, pH 6.8–7.4, *see* **Note 1**) glutaraldehyde (1%–3%) or a mixture of 2.5% glutaraldehyde and 4% formaldehyde in the

same buffer *(10)*. Light microscopy fixatives, such as AFA, Bouin's, will also produce adequate fixation for routine SEM.

3. Buffers such as sodium or potassium phosphate and sodium cacodylate at a concentration of 0.1 *M* (*see* **Note 2**).
4. Secondary fixative of 1% osmium tetroxide (0.1 *M*, pH 7.2, *see* **Note 3**).
5. Dehydrating agents, which are a graded series of either ethanol/water or acetone/water mixtures. A suitable series contains the following percentages of the organic component by volume: 30%, 50%, 70%, 80%, 90%, 95%, 100% (anhydrous) (*see* **Note 4**).
6. Dry siphoned liquid CO_2 cylinder (*see* **Note 5**).
7. A critical point dryer. Several types are available, but the most commonly used one is the Polaron (**Fig. 1**), which has to be connected to a CO_2 cylinder, hot and cold water supply, drain, and exhaust tube (*see* **Note 6**).
8. Safety equipment: gloves, face shield, fume hood, metal mirror, well-ventilated room, ear plugs (*see* **Note 7**).
9. Desiccant container for dried specimens (*see* **Note 8**).
10. Items for sample preparation, which may include filters, a filter container and syringe, small round glass cover slips coated with a cationic molecular adhesive such as poly-L-lysine *(11)*, porous containers, or glass vials (*see* **Fig. 2**).
11. Spare window seals for the critical point dryer. Dowty seals are used for the Polaron dryer.

3. Method

3.1. Specimen Preparation

1. Very small specimens (<100 µm). This group includes specimens such as bacteria, single cells or cultured cells, most protozoa and spores. Because of their size, these samples must first be attached to substrates that can then be processed through all stages. A common method is to filter suspensions onto 1-cm diameter Millipore, or polycarbonate filters using a Swinnex filter container and syringe. Polycarbonate filters are preferable, as they have a smooth and uniform background *(9)*. Alternatively, if the specimen surface carries a net negative charge, small, round, glass coverslips coated with a cationic molecular adhesive such as poly-L-lysine can be used *(11)*. Tissue cultures can be grown directly on coverslips.
2. Small specimens (100–500 µm). Specimens in this size range, i.e., small invertebrates, can be processed from fixation through dehydration using porous containers (**Fig. 2**) that can be quickly fabricated from BEEM capsules *(6)* or obtained through electron microscopy supply companies. These should be small enough to fit into the specimen carrier of the CPD apparatus.
3. Medium to large specimens (>0.5 mm). Specimens that are easily manipulated with forceps, i.e., large invertebrates, tissue blocks, and so on, can be processed using glass vials (**Fig. 2**) and then transferred to the metal carrier baskets that come with the CPD apparatus.

3.2. Processing

1. Immerse individual specimens, porous vials, filters or cover slips into fixative for the appropriate length of time (*see* **Note 9**).
2. Rinse briefly in two changes (5–10 min each) of the same buffer that was used to prepare the fixative.
3. Postfix for 1 to 2 h in osmium tetroxide solution (*see* **Note 10**).
4. Dehydrate with ethanol or acetone. The total time is determined by specimen size; small specimens require only a single change of 3 to 5 min in each concentration, while larger specimens may require two 10-min changes per step.
5. Precool the CPD apparatus to around 10°C by running cold water through the water jacket (*see* **Note 11**).
6. Transfer specimens from the 100% dehydrating solution into the sample carrier (*see* **Note 12**). Insure that the CO_2 inlet valve on the CPD apparatus is closed and then turn on the valve at the top of the CO_2 cylinder.
7. Unscrew the door of the CPD apparatus, load the sample trough, and then screw the door back on.
8. Open the CO_2 inlet valve two or more turns and then open the vent valve on top of the CPD apparatus slightly to avoid back-pressure and allow a quick fill.
9. While keeping the vent valve slightly open to allow CO_2 gas to escape, open the drain valve at the bottom of the apparatus for about 15 to 30 s to flush away most of the dehydrating fluid through the exhaust tube (*see* **Note 13**).
10. Flush the apparatus as outlined above for 1 to 2 min every 0.5 h for 0.5 to 3 h depending on specimen size (*see* **Note 13**). Remember to leave the room door wide open to dissipate the CO_2 gas being vented (*see* **Note 6**).
11. After flushing, close the inlet valve on the CPD apparatus and turn off the valve on the CO_2 cylinder. Next, lower the liquid level to just below the top of the specimen carrier by venting off excess CO_2 through the vent valve.
12. Run hot water through the water jacket while monitoring the temperature and pressure on the CPD gauges. When the pressure rises to 8.3 Pa (1200 psi) or slightly above, and the temperature reaches about 36 to 38°C, the liquid/gas boundary line will disappear indicating that the CO_2 fluid in the chamber is above the critical point.
13. Once the critical point has been exceeded, vent the gas off slowly (over a 5- to 10-min period) to avoid condensation of the CO_2 (*see* **Note 14**). The temperature gauge should remain between 35°C and 45°C during this time.
14. Shut off the water, unscrew the door, and remove the sample carrier with the dried specimens (*see* **Note 8**). It is sometimes helpful to sniff the carrier immediately upon removal to confirm that no trace of ethanol remains.
15. Occasional maintenance of the apparatus may be required (*see* **Note 15**).

4. Notes

1. Mature plant tissue generally requires a pH below 7.0 and more concentrated fixatives to buffer vacuolar contents ***(12)***.

2. Phosphate buffers are considered to be more physiologically compatible with most specimens, but have a shorter shelf-life due to microorganism contamination. Cacodylate buffers are toxic, but are stable for months.
3. Secondary fixation in 1% osmium tetroxide in the same buffer used for the primary fixation will further fix the specimen, but can also result in some extraction of proteins necessary for structural integrity. A variant of osmium fixation involves the use of thiocarbohydrazide as a ligand that can bond to additional osmium tetroxide. This method, called the OTO procedure *(13)*, is believed to strengthen the specimen and render it electrically conductive so that signal strength is improved and heating is minimized when the specimen is scanned with the SEM beam.
4. Ethanol has less potential for extraction of tissue components, but is not as miscible with CO_2 as acetone. Acetone can be obtained without a license and is more easily kept anhydrous. 100% solutions of either reagent should be stored over a desiccant such as calcium aluminosilicate pellets (available from electron microscopy supply companies) to maintain dryness. To avoid contamination of the specimen with particles that originate from the pellets, they can be placed in dialysis tubing before adding to the reagent bottle. Both reagents cause some specimen shrinkage, but more has been reported for acetone. By gradually increasing the concentration of the dehydrating agent from 70% to 100%, shrinkage is minimized.
5. A CO_2 cylinder with a siphon tube is required to supply liquid CO_2 to the CPD apparatus. Since CO_2 is supplied in several grades, it is also important to specify that you require extra dry when ordering the cylinder. If dry CO_2 is not obtainable, a high-pressure filter containing a molecular sieve can be placed in the line to remove moisture, and replaced when cylinders are changed. It is also essential to anchor the cylinder to the wall.
6. A long plastic tube can be attached to the drain vent so that cold vapors and frozen CO_2 can be exhausted safely. Ideally, the exhaust tube should be vented into a fume hood to prevent elevated CO_2 levels in the room.
7. Gloves and a fume hood should be used when working with fixatives, particularly osmium tetroxide. A metal mirror on a stand alleviates the necessity of looking directly into the window of the CPD apparatus, and a face shield can be used for protection as well.
8. If dried specimens are not mounted on stubs directly, they should be stored desiccated to prevent shrinkage and swelling artifacts that can occur due to humidity changes. Specimens can be stored in porous capsules in larger glass scintillation vials that have a capsule of desiccant (available from EM supply companies) glued to the lid (**Fig. 2**). All specimens, mounted or unmounted, should be stored desiccated to preserve structural integrity.
9. Cover slip and filter paper samples require only 0.5 h to fix completely, while unmounted individual specimens may require up to 3 h depending on size. It is generally believed that at least one dimension should be 1 mm or less to ensure adequate penetration of the fixatives. Short-term storage (up to 16 d)

in glutaraldehyde apparently produces no deleterious effects on membrane surfaces, particularly if postfixation in osmium follows. Long-term storage in glutaraldehyde, however, has been shown to cause perforations in membrane surfaces, even if postfixation in osmium is not used ***(14)***.

10. Prolonged fixation in osmium can cause extraction of membrane lipids. It is also possible to fix some specimens directly in osmium vapor ***(9)***, a method that has produced excellent results with fungal specimens.
11. Precooling the CPD apparatus ensures that CO_2 liquid rather than gas will fill the pressure chamber initially. Cooling too much, however, can lead to moisture condensation in the chamber when the door is opened. In high humidity environments, therefore, it may not be desirable to precool the apparatus.
12. Take care to prevent the tissue samples from drying prematurely during transfer, particularly if acetone is being used. The less dehydrant used, the better, however, as more time and liquid CO_2 are required to flush the excess away. Add only enough to partially cover the specimens as wicking will keep the tops wet. If cover slips are used, make sure they are completely covered with dehydrating fluid at all times.
13. It takes some practice to be able to admit liquid CO_2 quickly so that the specimens do not dry prematurely. The key is the correct combination of gas and liquid venting, which is achieved by judicious operation of the vent and drain valves. Creating turbulence in the pressure chamber by rapidly cycling the venting of liquid CO_2 will aid in the mixing and removal of dehydrating fluid. This can be done during each flush, but it is important to maintain the liquid CO_2 level above the tops of the specimens to prevent premature drying. Care must be taken not to freeze the drain valve during prolonged flushing. If freezing is a recurrent problem, it may be prudent to keep a hair dryer handy to thaw out the valve if necessary. It is also advisable to wear earplugs during this operation, as the escaping CO_2 can be very loud.
14. Slow venting at 100–200 psi/min will prevent retrograde condensation, or condensing of vapor within the pressure chamber. This will only happen if all of the dehydrating solvent has not been removed with flushing.
15. With respect to maintenance, it is recommended that the door seal, screw thread, and seals, and O-rings of the drain valve be cleaned of hard particles (e.g., shards of glass from coverslips) periodically to avoid scoring of these parts. This can be done with a cotton swab moistened with ethanol or acetone. Occasionally, the nitrile component of the Dowty seal on the window becomes cracked and leaks. When this happens, it has to be replaced by removing the perspex shield with the Allen wrench provided and unscrewing the metal retainer ring that holds the window in place (a tool for this is also supplied with the apparatus). It is advisable therefore to have one or more spare Dowty seals on hand.

References

1. Anderson, T. F. (1951) Techniques for the preservation of three-dimensional structure in preparing specimens for the electron microscope. *Trans. N.Y. Acad. Sci.* **13,** 130–133.

2. Oster, G. and Pollister, A. (1966) *Physical Techniques in Biological Research*, 2nd ed., Vol. 3, Academic Press, New York, p. 319.
3. Bray, D. F., Bagu, J., and Koegler, P. (1993) Comparison of hexamethyldisilazane (HMDS), Peldri II, and critical point drying methods for scanning electron microscopy of biological specimens. *Micros. Res. Tech.* **26,** 489–495.
4. Dey, S., Basu Baul, T. S., Roy, B., and Dey, D. (1989) A new rapid method of air-drying for scanning electron microscopy using tetramethylsilane. *J. Microsc.* **156,** 259–261.
5. Bozzola, J. J. and Russell, L. D. (1992) *Electron Microscopy: Principles and Techniques for Biologists*, Jones and Bartlett, Boston, MA, pp. 40–53.
6. Newell, D. G. and Roath, S. (1975) A container for processing small volumes of cell suspensions for critical point drying. *J. Microsc.* **104,** 321–323.
7. Hayat, M. A. (1978) *Introduction to Biological Scanning Electron Microscopy*, University Park Press, Baltimore, MD, pp. 150–162.
8. Cohen, A. L. (1979) Critical point drying-principles and procedures. *Scanning Electron Microsc.* **II,** 303–323.
9. Watson, L. P., McKee, A. E., and Merrell, B. R. (1980) Preparation of microbiological specimens for scanning electron microscopy. *Scanning Electron Microsc.* **II,** 45–56.
10. Karnovsky, M. J. (1965) A formaldehyde-glutaraldehyde fixative of high osmolarity for use in electron microscopy. *J. Cell Biol.* **27,** 137a.
11. Mazia, D., Sale, W. S., and Schatten, G. (1974) Polylysine as an adhesive for electron microscopy. *J. Cell Biol.* **63,** 212a.
12. Falk, R. H. (1980) Preparation of plant tissues for SEM. *Scanning Electron Microsc.* **II,** 79–87.
13. Kelley, R. O., Dekker, R. A. F., and Bluemink, J. G. (1975) Thiocarbohydrazide-mediated binding: a technique for protecting soft biological specimens in the scanning electron microscope, in *Principles and Techniques of Electron Microscopy,* Vol. 4 (Hayat, M. A., ed.), Van Nostrand Reinhold, New York, pp. 34–44.
14. Boyde, A. and Maconnachie, E. (1981) Morphological correlations with dimensional change during SEM specimen preparation. *Scanning Electron Microsc.* **IV,** 27–34.

32

Staining of Fingerprints on Checks and Banknotes Using Ninhydrin

Anthony A. Clifford and Ricky L. Green

1. Introduction

Latent fingerprints on paper and other porous surfaces can be developed using chemical methods so that they become visible to the naked eye and available as forensic evidence *(1,2)*. Ninhydrin is the most commonly used reagent for developing fingerprints as it reacts with amino acids present in ecrine sweat to give the strong purple color familiar when it is used as a stain for protein. Another compound used is 1,8-diazafluorene-9-one (DFO), which gives fingerprints that fluoresce, and is claimed to be more sensitive. Paper evidence, such as checks and banknotes, are treated by immersing the paper in a tray of a solution of the reagents and allowing to dry. The solution can also be brushed onto cardboard or wallpaper. The latent fingerprints are then developed by heating the paper in a specially adapted oven at 80°C and 65% relative humidity. DFO-treated surfaces, however, are treated at 100°C with no added humidity *(3)*.

Initially, many of the solvents used were highly flammable, which presented a hazard. For this reason, more recently trichlorotrifluoroethane (CFC113) is often used, and it has further advantages of being very volatile and not causing diffusion of handwriting. However, CFC113 is an ozone-depleting substance and therefore is being phased out *(4)*. As part of the search for a replacement, experiments have been carried out by using supercritical carbon dioxide *(5)*. These showed that fingerprint development using ninhydrin could be successfully carried out in the medium. Furthermore, development occurred in a one-step process, and it was not necessary to carry out a second stage in an oven, as is the case when liquid solvents were used. It was also shown that pure carbon dioxide and carbon dioxide modified with 5% methanol by volume did not cause printing and handwriting, either from ink or a ballpoint pen, to diffuse.

From: *Methods in Biotechnology, Vol. 13: Supercritical Fluid Methods and Protocols*
Edited by: J. R. Williams and A. A. Clifford © Humana Press Inc., Totowa, NJ

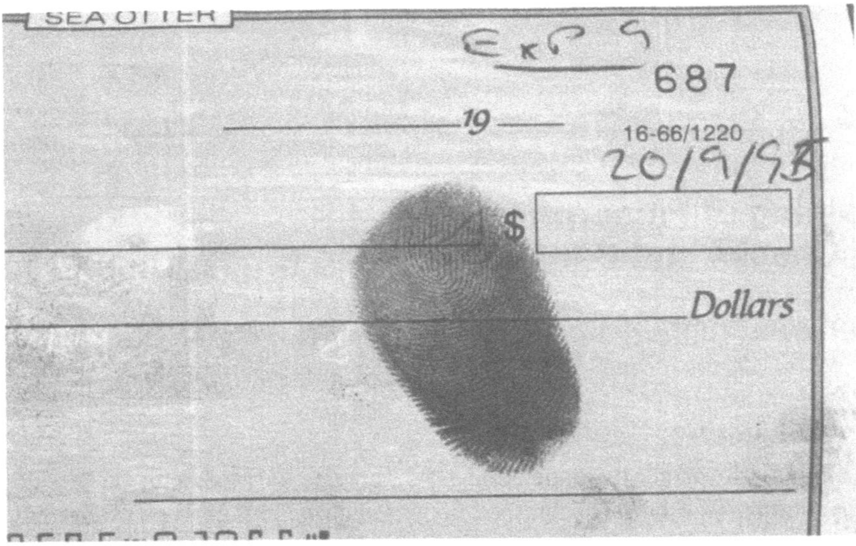

Fig. 1. Fingerprint developed on a check using ninhydrin in carbon dioxide.

The amino acids serine and glycine were shown not to dissolve in carbon dioxide, even when containing 5% methanol by volume.

This chapter describes the staining of fingerprints using the reaction between ninhydrin and amino acids in supercritical carbon dioxide. The work reported was aimed at a possible replacement of the solvent methods currently used. Ninhydrin is established as a reagent for staining amino acids and proteins in various applications. The method described here may therefore have other biological applications in locating proteins and amino acids. The development of "sweaty" fingerprints is successful under mild conditions and that of "greasy" fingerprints can also be achieved under more stringent conditions. A number of checks can be treated simultaneously under these more stringent conditions. An example of a successfully stained fingerprint is shown in **Fig. 1**. Similar procedures are likely to be successful with DFO, if sufficient methanol is used, as it was shown that DFO can be eluted through a chromatographic column when carbon dioxide containing 5% methanol by volume is used *(5)*.

2. Materials

1. A laboratory-scale high-pressure system as illustrated in **Fig. 2**. The system should include a cell, in which the reagents and the sample to be stained is placed, and which should have an easily removable lid. The cell should be placed in a heater capable of controlling the temperature of the cell to ±3°C between 40°C and 100°C. There should be a side-arm connecting the top and bottom of the cell, which is outside the heater and, therefore, at room temperature. This arrange-

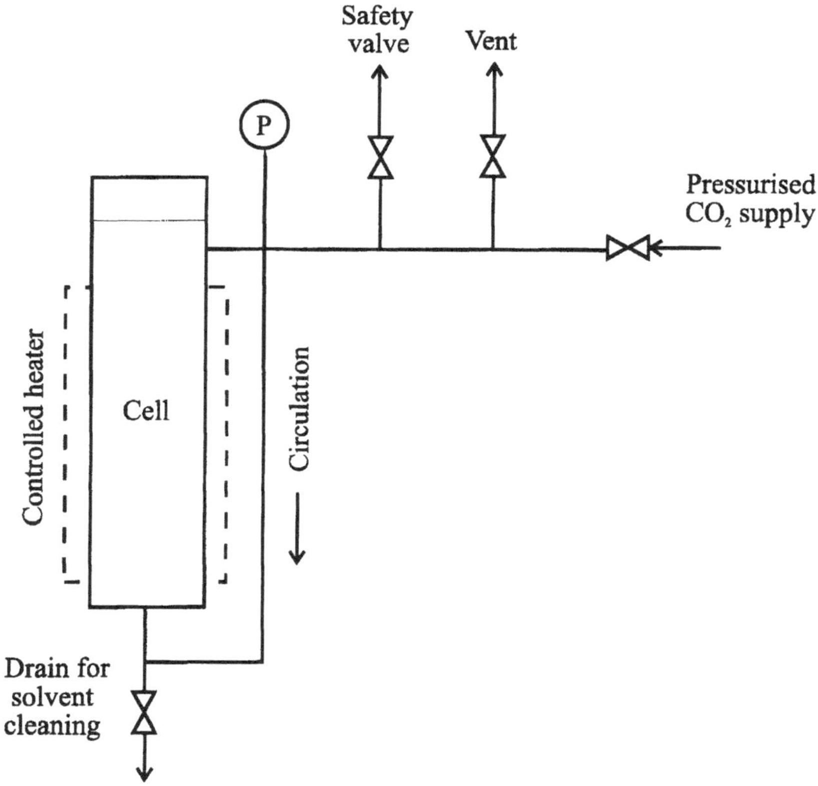

Fig. 2. A laboratory-scale system for the development of fingerprints using supercritical carbon dioxide.

ment causes circulation of the fluid down through the side-arm and up through the cell because of the higher density of the fluid in the side-arm. Other required features of the system are a pressure gauge and a spring-operated safety pressure relief valve, which opens if the pressure exceeds 320 bar. Also needed are connections via valves for filling with pressurized carbon dioxide, for venting the system at the end of the procedure and for allowing cleaning of the system with solvent. The vents from the venting and pressure relief valves should be led outside the building or into a fume cupboard. The system described can be made by adapting a commercial supercritical fluid extraction system, which is available either on a laboratory or pilot plant scale, or constructing the system in-house from available components (*see* **Note 1**). The system should be pressure-tested (*see* **Note 2**).

2. A pump capable of delivering liquid carbon dioxide to a pressure of 300 bar and at a flow rate of 10 mL/min (*see* **Note 3**).
3. A cooler for the pump-head (*see* **Note 4**).

4. A cylinder of industrial-grade carbon dioxide (98.5%) fitted with a diptube (*see* **Note 4**).
5. Aluminum foil.
6. Staining reagent, which is made up by dissolving 3.0 g of ninhydrin in 10 mL of a 50% mixture of water and acetic acid by volume.
7. A supply of spare "O" rings for the quick-release cap.
8. Rubber gloves, for use when handling the sample.

3. Method

1. The staining reagent (0.1 mL) is placed in a small tray made out of aluminum foil in the bottom of the cell.
2. With gloved hands, the paper on which fingerprints are developed, for example, checks or banknotes, are rolled up and pushed into the cell and are thus lying against the inner walls.
3. The cell is then sealed.
4. The temperature is set to 80°C (*see* **Note 5**).
5. The system is pressurized with carbon dioxide to 250 bar over a period of 5 min (*see* **Note 5**). Pressure adjustment may be required over the next 5 min by pumping in more carbon dioxide or releasing it.
6. The system is then left under the same temperature and pressure for a further 30 min.
7. The heater is then turned off.
8. The pressure is then released over 30 min (*see* **Note 6**).
9. The system is then left for a further 10 min at atmospheric pressure to ensure that all the carbon dioxide has escaped and desorbed from the "O" ring.
10. The quick-release cap is removed and the sample removed for examination.

4. Notes

1. The system used in our laboratory is built as follows and has also been used to carry out the first experiments in which cotton was successfully dyed from carbon dioxide *(6)*. The cell is machined out of 316 stainless steel and fitted with a quick-release cap, which is made from a quick-release connector and a stop-end connector. It is 100 mm long and has a volume of 50 mL. It is installed in a temperature-controlled heater, built in-house. The heater consists of resistive wire wound on to a ceramic tube, thermally and electrically insulated and placed in an aluminum box. The tube also contains a thermocouple which feeds the heater controller. The tubing used is .025-inch stainless steel and the components used to assemble the system, including the connection used to make the quick-release cell cap, are supplied by the Manchester Valve and Fitting Company, Manchester, UK. The pressure gauge is a Bourdon Gauge supplied by Buddenberg, Manchester, UK. The system is designed for a maximum working pressure of 300 bar at 100°C and the pressure release valve was set to 320 bar.
2. The system should be checked for leaks and the ability to withstand pressure, whether a commercial system is adapted or the whole system built in-house. First, it should be checked for leaks using nitrogen at five bar and a solution of deter-

gent. Then, with the pressure release valve removed, the system should be filled with water, taking care to flush out all air. It should then be pressurized behind a safety screen to 450 bar by pumping in water from a high performance liquid chromatography (HPLC) pump. The water pump should then be isolated by a valve and the pressure observed over 1 h. The pressure may fluctuate because of temperature changes, but should not consistently fall.

3. A wide range of pumps designed for HPLC are suitable. If pump-head cooling is to be used (*see* **Note 4**), the pump-head must be accessible.
4. Carbon dioxide must be pumped as a liquid and there are two ways of achieving this. The first option is to cool the pump-head to a maximum of 5°C by using a pump-head jacket, which can be built in-house, through which an antifreeze solution at about –5°C is pumped from a laboratory chiller. In this case, industrial-grade carbon dioxide can be used, which is relatively inexpensive. Alternatively, pump-head cooling is dispensed with and a carbon dioxide cylinder with a helium overpressure is used. In these cylinders, which are expensive and supplied mainly for supercritical fluid chromatography, the helium pressure above the carbon dioxide is above 100 bar, ensuring that the carbon dioxide is liquid in the pump-head at room temperature and a few degrees above.
5. These conditions will cope with both "sweaty" and "greasy" fingerprints on five checks or banknotes. For a single check with a "sweaty" fingerprint, a temperature of 40°C and a pressure of 125 bar are sufficient.
6. The "O" rings in the quick-release cap absorb carbon dioxide considerably and swell. If the pressure is released quickly they will explode. Pressure is therefore released slowly, especially toward atmospheric pressure, over a period of 30 min. With careful use, the "O" rings can be reused up to five times, although they are subject to blistering.

References

1. Lee, H. C. and Gaensslen, R. E. (1991) *Advances in Fingerprint Technology*. Elsevier, New York.
2. Kent, T. (1992) *Manual of Fingerprint Development Techniques*. British Home Office, London.
3. Hardwick, S., Kent, T., Sears, V., and Winfield, P. (1993) Improvements to the formulation of DFO and the effects of heat on the reaction with latent fingerprints. *Fingerprint World* **19,** 65–69.
4. Dalyell, T. (1995) On the trail of the green fingerprint. *New Scientist* April 1st, 51.
5. Hewlett, D. F., Winfield, P. G. R., and Clifford, A. A. (1996) The ninhydrin process in supercritical carbon dioxide. *J. Forensic Sci.* **41,** 487–489.
6. Özcan, A. S., Clifford, A. A., and Bartle, K. D. (1998) Dyeing of cotton fibres with disperse dyes in supercritical carbon dioxide. *Dyes Pigments* **36,** 103–110.

Index

FSC
www.fsc.org
MIX
Papier | Fördert gute Waldnutzung
FSC® C083411

Kolibri 360 GmbH
Ferdinand-Jühlke-Straße 7
99095 Erfurt, Deutschland
produktsicherheit@kolibri360.de